Supervised Machine Learning for Science

How to stop worrying and love your black box

Christoph Molnar & Timo Freiesleben

Supervised Machine Learning for Science

How to stop worrying and love your black box

ISBN 978-3-911578-00-4 (paperback)
ISBN 978-3-911578-01-1 (hardcover)

2024, First Edition

Publisher: Self-published

Christoph Molnar
c/o DRA
Bayerstraße 77c
80335 München, Germany

commit id: 03de158

Content

Summary

Machine learning has revolutionized science, from folding proteins and predicting tornadoes to studying human nature. While science has always had an intimate relationship with prediction, machine learning amplified this focus. But can this hyper-focus on prediction be justified? Can a machine learning model be part of a scientific model? Or are we on the wrong track?

In this book, we explore and justify supervised machine learning in science. However, a naive application of supervised learning won't get you far because machine learning in raw form is unsuitable for science. After all, it lacks interpretability, causality, uncertainty quantification, and many more desirable attributes. Yet, we already have all the puzzle pieces needed to improve machine learning, from incorporating domain knowledge to creating robust, interpretable, and causal models. The problem is that the solutions are scattered everywhere.

In this book, we bring together the philosophical justification and the solutions that make supervised machine learning a powerful tool for science.

The book consists of two parts:

- Part 1 justifies the use of machine learning in science.
- Part 2 discusses how to integrate machine learning into science.

Preface

We wrote this book because we are passionate about science and machine learning, particularly their interaction.

Our collaboration began during our Ph.D. studies, where we co-authored papers on interpretable machine learning. However, we soon realized that interpretability was but one piece of a puzzle. Exploring causality, uncertainty quantification, robustness, and other tools proved essential to convert 'raw' machine learning into a proper scientific tool. The deeper we dug, the clearer it became that an even bigger piece was missing: A strong justification for using machine learning in scientific modeling was needed. All our insights led us to this book.

Prepare for a journey through the philosophy of science, machine learning theory, pragmatic modeling advice, and short stories of our raven scientists.

1. Introduction

Researchers make millions of protons collide every second inside the Large Hadron Collider (LHC) deep below the French-Swiss border near Geneva to understand what our universe is ultimately made of. Each collision creates subatomic particles which often decay quickly, creating even more particles. Sensors surround the collision zone and record the passage of these particles, turning each collision event into data – a lot of data. CERN, the European Organization for Nuclear Research, is the world's largest particle physics laboratory. CERN's data center processes up to a petabyte of data per day. That's a million gigabytes. These amounts of data are far beyond anything a human can look at.

To call the analysis of this data "challenging" is a complete understatement: It's difficult to reconstruct the collisions because multiple collisions occur simultaneously; sensor data is generated faster than it can be written to disk; and some particles decay in less than a trillionth of a trillionth of a second. To make the data processing and analysis manageable, CERN researchers rely on machine learning. Machine learning systems scan the incoming event data and decide which to write to disk and which to ignore. Machine learning models also estimate the energy and timing of signals. Other machine learning models help reconstruct events and remove noise. Today, particle science at CERN is unthinkable without machine learning.

But it's not just particle science. If you look closely, you'll find machine learning in almost every field (at least, we couldn't find any without it). There are fields where everyone expects machine learning applications, such as geoscience, materials science, or neuroscience [1], [2], [3]. But you might be surprised to see machine learning applied in other fields, such as anthropology, history, or theoretical physics [4], [5], [6].

Let's look at a few applications to get a better sense of how scientists use machine learning. In each example, machine learning plays a different scientific role.

Raven Science was stuck. The Ravens were drowning in complex data and had no idea how to compress it into a scientific model. A peculiar bunch of neuro-stats-computer Ravens had recently developed a new modeling approach called supervised machine learning, which learns predictive models directly from data. But is this kind of modeling still science, or is it alchemy?

1.1. Labeling wildlife in the Serengeti

Wildlife ecologists want to understand and control the complexities of natural ecosystems. They ask questions like: What animals live in the ecosystem? How many animals of each species are there? What do the animals do?

For a long time, it was difficult to analyze these questions quantitatively because there was simply no data. But today, motion sensor cameras, often referred to as *camera traps*, make huge amounts of data available. Camera traps combine cameras with motion sensors. Once something moves in front of the camera, it takes pictures. The Serengeti, home to zebras, buffalos, and lions, is one of the ecosystems where ecologists placed various camera traps. But to turn these images into insights, people had to label them: they had to decide whether an animal was in the picture, what species it belonged to, and describe what it was doing. A tedious task!

For this reason, Norouzzadeh et al. [7] used the already labeled data to train a convolutional neural network that performs this task automatically. Figure 1.1 shows the output of this labeling task. The model achieves a prediction performance of 94.9% accuracy on the Snapshot Serengeti dataset – the same performance as crowdsourced teams of volunteers who typically label the data.

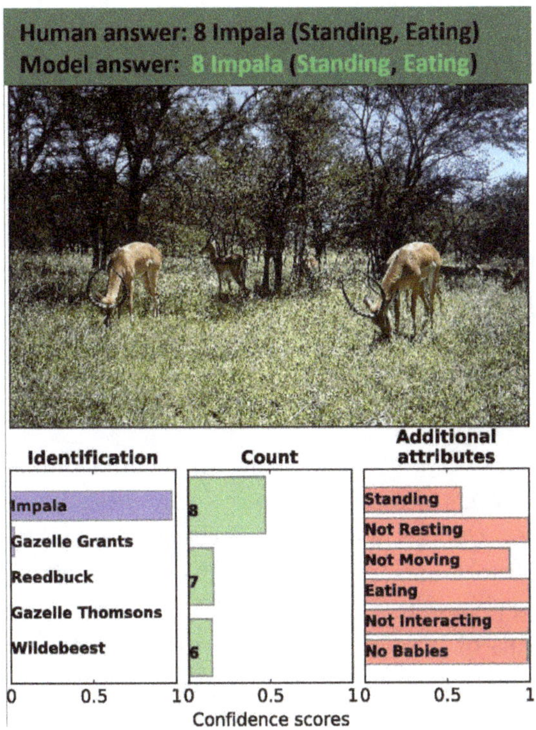

Figure 1.1.: The image was taken by motion sensor cameras in the Serengeti and shows wildlife animals. The machine learning model by Norouzzadeh et al. [7] correctly identifies, counts, and describes the animals in the image. Used with permission from [7].

The primary goal of the paper is to provide a reliable labeling tool for researchers. The labels for the data should be accurate enough to draw scientific conclusions. Therefore, Noroyzzadeh et al. [7] are concerned with the predictive accuracy of the model under changing natural conditions and the uncertainty associated with the predictions.

1.2. Forecasting tornadoes to inform actions

In the case of severe weather events like tornadoes, every minute counts. You need to find shelter for yourself and your loved ones quickly. But tornadoes are hard to predict. They form rapidly and the exact conditions for a tornado to form are not fully understood.

Lagerquist et al. [8] therefore use machine learning to predict the occurrence of a tornado within the next hour. They trained a convolutional neural network on data from two different sources: storm-centered radar images and short-range soundings. Their model achieves similar performance to ProbSevere, an advanced machine learning system for severe weather forecasting that is already in operation.

The work aims to provide a system that could be used in deployment to help warn the population of tornadoes. Therefore, the paper centers on the methodology and the assessment of the model's predictive performance in relevant deployment settings. In particular, they analyzed in which regions the model worked best and worst in terms of prediction error for different types of tornadoes. The use case differs from the Serengeti animal classifier where the goal was labeling, not case-by-case decision-making (although summary data from Serengeti might end up for decision-making).

1.3. Predicting almond yield to gain insights

California is the almond producer of the world: it produces 80% of all the almonds on Earth, probably in the entire Universe. Nitrogen, a fertilizer, plays a key role in growing the nuts.[1] Fertiliser is regulated, meaning there's an upper limit for its use. Before you can develop a good fertilizer strategy, you have to quantify its effect first. Zhang et al. [9] used machine learning to predict the almond yield of orchard fields, based on weather, orchard characteristics using remote sensing (satellites), and, of course, fertilizer use. The authors examined the model not only for its predictive accuracy but also for how individual features affect the prediction and how important they are for the model's performance.

The goal of this research is twofold: prediction and insight, so it goes beyond just making a decision. The model may help almond growers make better decisions about fertilizer use but also contribute to the scientific knowledge base. In general, the scientific fields of ecology and agriculture are at the forefront when it comes to adapting machine learning [10], [11].

1.4. Inferring protein structures for scientific hypotheses

One of the big goals in bioinformatics is to understand how protein structures are determined by their amino acid sequences. Proteins do a lot of heavy lifting in the body, from building muscles and repairing the body to breaking down food and sending messages. A protein's structure is largely determined by its sequence of amino acids, the building blocks of protein. The protein's structure, however, ultimately determines its function. The problem: It is difficult to predict how a protein will be structured if you only have the amino acid sequence. If scientists could do that reliably, it would help them with drug discovery, understanding disease mechanisms, and designing new proteins. Meet AlphaFold [12], a deep neural network that can predict protein structures from their amino acid sequence reasonably well. See Figure 1.2 for one example.

[1] Technically, they are legumes, but we're not biologists and we eat them as if they were nuts.

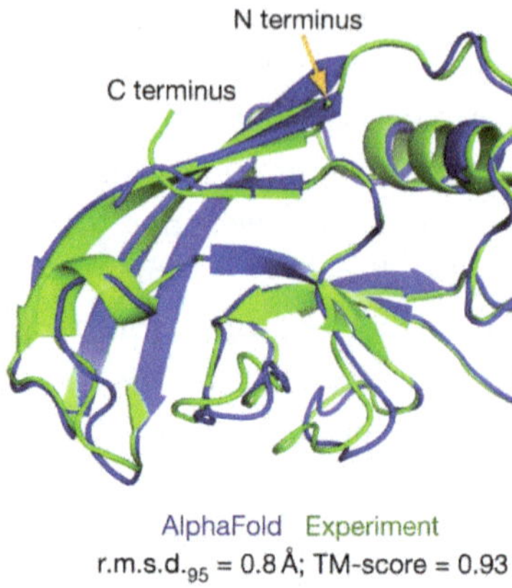

Figure 1.2.: Protein structure, predicted in blue, determined by experiment in green. Figure by [12], CC-BY (https://creativecommons.org/licenses/by/4.0/)

AlphaFold was trained on a dataset of 100,000 protein sequences. Beyond pure prediction, the algorithm aids with molecular replacement [13] and interpreting images of proteins taken with special microscopes [14]. There's now even an entire database, called AlphaFold DB, which stores the predicted protein structures. On its website [15] it says:

> "AlphaFold DB provides open access to over 200 million protein structure predictions to accelerate scientific research."

Not only has the prediction model AlphaFold become part of science, but its predictions serve as a building block for further research. To justify this sensitive role of machine learning, the model needs to be doubly scrutinized.

1.5. What role does machine learning play in science?

You may look at these examples and argue that machine learning is only one of many tools in the scientific process, like Excel or Python, or that the scientific goal was only to evaluate the model's performance.

If this were the case, the value of machine learning in science would be very limited – machine learning could create 'models', but these bear little resemblance to scientific models that represent phenomena. The most important methodological steps in science – such as creating scientific models or theories, testing them, and finding interesting new research problems – would remain unaffected by machine learning.

Indeed, there are good reasons why everyone is cautious about machine learning in science – in its raw form, machine learning reduces every scientific question to a prediction problem. This restraint is emphasized by popular critical voices about machine learning.

Take Judea Pearl, a proponent of causal inference, who said that

> "I view machine learning as a tool to get us from data to probabilities. But then we still have to make two extra steps to go from probabilities into real understanding — two big steps. One is to predict the effect of actions, and the second is counterfactual imagination." [16]

Pearl refers here to his ladder of causation that distinguishes three ranks: 1. association, 2. intervention, and 3. counterfactual reasoning [17]. He highlights that machine learning remains on rank one of this ladder and is only suitable for static prediction.

Gary Marcus, one of the most well-known critics of current deep learning claimed that

> "In my judgement, we are unlikely to resolve any of our greatest immediate concerns about AI if we don't change course. The current paradigm – long on data,

> but short on knowledge, reasoning and cognitive models – simply isn't getting us to AI we can trust." [18]

In the paper, Marcus criticizes the lack of robustness, the ignorance of background knowledge, and the opacity in the current machine learning approach.

Noam Chomsky, one of the founding fathers of modern linguistics argued:

> "Perversely, some machine learning enthusiasts seem to be proud that their creations can generate correct "scientific" predictions (say, about the motion of physical bodies) without making use of explanations (involving, say, Newton's laws of motion and universal gravitation). But this kind of prediction, even when successful, is pseudoscience. While scientists certainly seek theories that have a high degree of empirical corroboration, as the philosopher Karl Popper noted, 'we do not seek highly probable theories but explanations; that is to say, powerful and highly improbable theories.' " [19]

Chomsky criticizes machine learning for its exclusive focus on prediction rather than developing theories and explanations. In the article, he is particularly doubtful that current large language models provide deep insight into human language.

1.6. Machine learning can be more than a tool

We share these critical views above to a certain degree. Machine learning algorithms have problems with being dumb curve-fitters that lack causality. Purely predictive models might make bad explanatory models. And machine learning doesn't produce scientific theories as we are used to.

Nevertheless, there are reasons for optimism that these issues can be addressed and that machine learning can play a big role in science. The roles we see machine learning play were already shining through some of the initial examples:

- **Inform actions:** Tornado forecasting is not just an intellectual forecasting exercise, but is crucial for taking the right measures to prevent harm.
- **Gain insights:** In the case of almond yield prediction, the goal wasn't only to build a prediction machine, but the researchers also extracted insights from the model.
- **Exploration:** AlphaFold isn't a mere proof of concept. For example, protein structure predictions are used by researchers to explore and test new drugs.

However, machine learning cannot fulfill these roles in its raw form. It must be equipped with "upgrades" such as (causal) domain knowledge, interpretability, uncertainty, and so on, to become a new approach for informing peoples' actions, extracting insights from data, or generating new hypotheses. We believe that fully upgraded supervised machine learning models have the potential to become a full-blown scientific methodology that helps scientists to better understand phenomena.

The upgrades we need are readily available. Every year, new subfields are emerging. If we think of machine learning not in isolation, but in conjunction with these upgrades, machine learning could provide great scientific value. Like a puzzle, we have to piece all of it together, starting with a justification of why a machine learning approach focused on prediction is a good core idea for research.

But before we can justify machine learning or start solving the puzzle, let's look at supervised machine learning in its raw form!

2. Bare-Bones Supervised Machine Learning

This chapter looks at supervised machine learning, stripping it to its bare-bones. Why supervised machine learning? Of course, the applications of unsupervised learning and reinforcement learning in science are no less interesting. However, after looking at many scientific applications, we realized that the primary goal was often to solve a supervised learning problem. Unsupervised and reinforcement learning techniques have often been used to support this goal, e.g. by providing powerful representations or fine-tuning models.

What is supervised machine learning about? Think back to the tornado prediction example from the introduction.
Supervised machine learning produces models that provide *output predictions*, for example, concerning the occurrence of a tornado in the next hour. To obtain these predictions, models must be fed with so-called *input feature values*, e.g. storm-centered radar images and short-range soundings. The *search for models* is done via a *learning algorithm* and a *labeled dataset*, consisting of *input-output pairs*, also called training data.

A young Raven called Rattle was the first to adopt supervised machine learning. At first, the other Raven scientists were skeptical. Too new. Unproven. Risky. Not the way of Raven Science. Nevertheless, Rattle began to explain to the first interested Ravens what machine learning was all about.

2.1. Describe the prediction task

To use supervised machine learning, you first have to translate your problem into a prediction task with the following ingredients:

- **Pick a target variable Y to predict.** For example, the occurrence of tornadoes in a 1-hour time window, coded as 0 (no tornado) and 1 (tornado).
- **Define the task.** The task is related to the target and can range from classification and regression to survival analysis and recommendation. Depending on how you frame the tornado prediction problem, you end up with different task types:
  - Will a tornado occur within the next hour? ⇒ classification (Tornado Y/N)
  - How many tornadoes will occur this year? ⇒ regression (0, 1, 2, ...)
  - How long until the next tornado occurs? ⇒ survival analysis (e.g. 2h)

- **Decide on the evaluation metric.** This metric defines what counts as a good or bad prediction. To classify tornadoes in 1-hour windows, you could use metrics such as F1 score[1] or accuracy.
- **Choose features X from which to predict the target.** Features could be hourly measurements from radar stations, like wind speeds and precipitation.

Once the task is defined, you need data.

2.2. Get the data

Machine learning requires data. Data points are represented as pairs of feature and target values $(x^{(i)}, y^{(i)})_{i=1,\ldots,n}$. In the tornado case, $x^{(145)}$ might be radar measurements from 10 AM to 11 AM on May 25th, 2021, in a specific 3km by 3km patch in the United States. Accordingly, $y^{(145)}$ would indicate whether a tornado occurred in this time and place.

After cleaning and pre-processing the data, it is typically randomly split into three subsets for machine learning:

- A training dataset, D_{train}, used to train the model.
- A validation dataset, D_{val}, used to validate modeling choices and selection.
- A testing dataset, D_{test}, used for final performance evaluation.

The simplest way is to randomly split your dataset into these three buckets. In reality, you have to adapt the splitting mechanism based on data and tasks. For example, time series and clustered data require splitting schemes that respect the data structure. For classification with imbalanced data, you might want to use mechanisms that ensure that the minority class occurs often enough in each split. Anyways, in a completely random 3-way split, our data point $(x^{(145)}, y^{(145)})$ would fall into one of these buckets.

[1]The F1 score is a metric that balances precision and recall, particularly useful for evaluating performance on imbalanced datasets.

2.3. Train the model

Training a machine learning model means running an algorithm that takes as input the training data and outputs the model. The model describes a function that outputs predictions based on input features.

The training requires making some choices:

1. Select a class of models F. For example, decision trees or neural networks. Usually, you have to specify this class further, e.g., by choosing the maximal depth of a tree or the specific architecture of the neural network. Radar readings might be on a spatial level, so you might pick a convolutional neural network such as ResNet for the tornado prediction task.
2. Choose a training algorithm I. The training algorithm takes the training data and produces a prediction model $\hat{f}$. For example, neural networks typically use stochastic gradient descent with backpropagation as the training algorithm. But you could also train a neural network using genetic algorithms.
3. Set hyperparameters. Hyperparameters control the training algorithm I and affect the models that the training produces. Some hyperparameters are related to the model class, like the number of layers in your neural net or the number of neighbors in the k-nearest-neighbors algorithm. Others are connected to the training algorithm, such as the learning rate or the batch size in stochastic gradient descent.

For example, when you train a convolutional neural network to predict tornadoes, you use stochastic gradient descent (I) with the training data D_{train}. To do this, you have to set the hyperparameters. After the training process, you get a trained CNN $\hat{f}$ from the model class of CNNs.

The training process seeks to produce a model $\hat{f}$ that makes a minimal error on the training data:

$$\epsilon_{train}(\hat{f}) := \frac{1}{|D_{train}|} \sum_{x,y \in D_{train}} L(\hat{f}(x), y),$$

where L is the loss function the model optimizes for. Some training algorithms can optimize arbitrary (well, some constraints remain) loss functions directly, like neural networks using gradient descent, while other training algorithms have built-in and sometimes less explicit losses they optimize for, like greedy splits in CART decision trees [20].

2.4. Validate modeling choices

Training doesn't guarantee that it will produce the best-performing model. For example, you might not have picked the best model class, because you have used linear regression but the best model for that task might be a tree ensemble. And even if you picked the best model class, you might have set the hyperparameters non-optimally. You need a procedure to both pick a model class and the hyperparameters. A naive approach would be to compute the evaluation metric for the training data, but this would be a bad idea since some models may overfit the training data.

> **i Overfitting**
>
> We say a model *overfits* when it has good performance on the training data but performs poorly with new unseen data from the same distribution. A model that overfits is good at reproducing the random errors in training data but fails to capture the general patterns.

You typically compute the evaluation metric using a separate validation dataset D_{val}. Since the model wasn't trained using the validation data, you get a fair assessment of its performance. With the validation data, you can compare multiple models and hyperparameter configurations and pick the best-performing one. This allows you to detect underfitting and overfitting and to guard against these problems by properly regularizing and tuning the model.

2.5. Evaluate the model on test data

How well would your final model perform? Unfortunately, you can't use the evaluation metrics you have computed from training and validation data. Both will be too optimistic since you already used the training data to train the model and the validation data to make modeling choices. Instead, you have to evaluate the model performance on test data D_{test}. This gives you a realistic estimate of the model performance.

2.6. And repeat: the role of resampling data

Having just one split into training, validation, and testing is not very data efficient. Typically, the data is split multiple times. A common technique is cross-validation, which splits the data into k different parts. Let's say you use $k = 10$. Nine out of ten folds might be used for training and validation, and the tenth for test data. You cycle through the folds so that each fold is used as test data once. This way, you always use "fresh" data for evaluating the model. Other sampling methods such as bootstrapping and subsampling can be used here as well. But even having multiple folds may not give stable results – would you generate the fold splitting again, you may get different estimates. So another established method is to repeat the sampling.

You have another choice to make, and that is how to split training and validation data into nine parts. You could either do a single split or do cross-validation again. Like in the movie *Inception*, which is about dreams within dreams, you go one level deeper and do cross-validation within cross-validation, a procedure called nested cross-validation [21]. The advantage of (nested) cross-validation is better estimates of the model performance, and better models since you use the data more efficiently.

2.7. Bare-bones machine learning can be automated

Once you have defined the prediction task and your data, the entire training process can be completely automated. The subfield of machine learning called AutoML aims to completely automate the machine learning training process and make machine learning engineers redundant [22]. Upload your data, pick a column as your target, and pick an evaluation metric. Click a button. And the machine does everything for you. Data splitting, hyperparameter tuning, model selection, model evaluation. We call this automatable practice of machine learning "bare-bones machine learning". The big question is: How does such an optimization-focused approach mix with a complex practice like science?

Part I.

Justifying Machine Learning For Science

In this first part of the book, we look at machine learning from a philosophical angle. How can machine learning help to advance science? After all, isn't machine learning only a tool for making better predictions?

In this part, we discuss, on the one hand, the importance that predictions take in science and justify machine learning based on its predictive powers. On the other hand, we emphasize the importance of other scientific goals like enabling control, explanations, and reasoning, and describe how they clash with the hyperfocus of current machine learning on prediction. We finish this part by highlighting several concrete limitations of bare-bones machine learning. To learn how to tackle these limitations, you should check out Part II of the book.

Concretely, Part I consists of the following chapters:

- Chapter 3 : Prediction takes a central role in science.
- Chapter 4 : Machine learning enhances scientists' predictive powers.
- Chapter 5 : Science has goals beyond prediction like control, explanation, and reasoning.
- Chapter 6 : Bare-bones supervised machine learning has concrete limitations.

3. The Role of Prediction in Science

Supervised machine learning is the discipline of producing high-performing prediction models. While the complex dance between prediction and science is age-old, supervised machine learning is relatively young. Let's find out how this algorithmic approach to learning from data fits into science.

> **Prediction**
>
> A prediction is an educated guess about an unknown outcome based on current information, such as predicting tomorrow's weather or the 3D structure of a protein from its amino acid sequences.

Rattle and a group of machine learning advocates presented a petition to the Elder Council of Raven Science. It contained a series of prediction problems, all of which could be solved by machine learning, such as predicting tornadoes or distinguishing poisonous berries from healthy ones. The Council began a difficult philosophical debate: What role should prediction play in Raven Science?

3.1. Scientific theories produce predictions

Since this book has "science" in the title, we are obliged to mention Einstein and his theory of general relativity, one of the biggest success stories of science. Inspired by theoretical and philosophical reflections such as the principle of relativity, Mach's principle, and the equivalence principle[1], Einstein described a theory according to which gravity is a geometric property of space and time. When Einstein developed his theory between 1907 and 1915, it was, unlike his theory of special relativity, not based on novel empirical observations that the old theory did not account for. Instead, Einstein was driven by intuition, a desire to generalize, and mathematical beauty. Still, Einstein's theory provided what elevated it beyond purely philosophical speculation: It made predictions.

Einstein's theory of general relativity predicted that:

- Black holes exist.
- Time slows down near massive objects.
- Gravity bends light, makes waves, and affects the red shift in light.
- Massive rotating objects can drag space-time with them, also called the Lense-Thirring precession.

All of these predictions have been confirmed empirically. Some predictions were confirmed soon after Einstein formulated his theory, such as that gravity bends light, which was observed by Eddington and Dyson in 1919 at a solar eclipse. Others, like the existence of gravitational waves, took until 2015 to be confirmed because they required complex new technologies of measurement. It is surprising how long Einstein's theory has lasted and how correct its predictions are until today – the general theory of relativity is among the best-tested theories in science.

[1]The principle of relativity says that physical laws are the same for all observers. Mach's principle states that objects move relative to each other, not in absolute space. The equivalence principle states that gravitational mass and inertial mass are equal.

Scientific Theories

Scientific theories are abstract and universal explanations for phenomena that have strong and diverse evidential support.[2] They are strongly tested and cohere with other theories. Theories are thus widely accepted in their fields. Furthermore, theories are not necessarily quantitative. Some of the most important scientific theories are qualitative, such as the germ theory of disease, Darwin's theory of evolution, and cell theory.

3.2. Prediction can be the goal of science

Less well-known than Einstein's story is the story of Lewis Fry Richardson, the first person to predict the weather using numerical methods. Besides meteorology, Richardson was an interesting and widely skilled character: he developed a sonar after the tragedy of the Titanic, he found inconsistencies in the measurement of the length of coasts that later inspired Mandelbrot for his theory of fractals, and he was a convinced pacifist and brought quantitative methods to study peace and conflicts between nations.

When Richardson started to work on meteorology, weather forecasting was largely based on experience with similar weather conditions. He wasn't the first to propose the use of physical theories in weather forecasting, but he was the first to test his numerical model in practice. He developed a mathematical model aimed at predicting the weather:

- He divided a weather map into squares of 200km side length.
- He subdivided each of the squares into layers according to their respective heights.
- He listed all the physical equations he deemed important to model the dynamics.

[2] In philosophy, as always, there are huge fights about what the terms theory, models, and hypothesis mean. As this is no encyclopedia article on the topic (check out [23]), we won't do justice to the intricacies of these debates.

- He solved the respective nonlinear partial differential equation to predict air pressure, wind speed, and temperature for two of the squares.

> **i** Scientific Models
>
> Scientific models are typically mathematical representations of aspects of phenomena that allow scientists to predict, explain, or reason about these phenomena. Models can be grounded in theory or even be applications of theories to a specific context. For instance, Richardson used theories from physics to build his weather forecasting model. However, models don't have to be closely tied to theory. They can also rely on data or common sense. Contrary to theories, models are often more concrete with a smaller scope and more practical. Evidence incompatible with the model just limits the scope of the model, but an otherwise useful model might still survive. Different models for similar aspects of reality can coexist in a field even if they are in conflict.

Unfortunately, his first attempts were unsuccessful. It took him **six weeks** to make a weather forecast for a **six-hour** time frame. Even worse, his predictions were pretty far off! Nevertheless, methodologically, he was on the right track; while first ignored after its publication in 1922, his book "Weather Prediction by Numerical Process" is now a modern classic in meteorology. Up until today, scientists use numerical methods, physical measurements, and physical theories to forecast the weather [24]. Fortunately, numerical weather forecasts have substantially improved since Richardson's first attempt. The availability of atmospheric data, e.g., via airplanes or weather stations, and even more importantly, the access to electronic computers that can solve differential equations efficiently in real-time, allow to predict the weather (at least within 24 hours) reliably. Weather forecasts have become so reliable that when you ask a friend about the weather at the moment, she may check her weather app instead of looking outside ...

3.3. Prediction and science are inseparable

Prediction was central to both Einstein's general theory of relativity and Richardson's weather forecasting. If the predictions of general relativity had been incorrect, the theory would have been thrown on the scrap heap of science, no matter how beautiful it was. Similarly, if numerical weather forecasting did not work, it would be ignored; in fact, this is exactly what initially happened to Richardson when he proposed his theory.

Predictions played different roles in both cases: In Einstein's case, the predictions guided new experiments such as the search for black holes, confirmed the theory, and had in the beginning very few practical implications.[3] In weather science, on the other hand, accurate prediction is ultimately the goal; also because controlling the weather is at least so far out of reach. Weather predictions are useful: They help people decide when to leave the house, when to harvest plants, how to schedule flights, when to expect catastrophic weather events, and tell your trusted ice shop when to open. Prediction is not a nice byproduct of weather science – it is the goal.

It is widely agreed that prediction is one of the core aims of science [25]. Prediction connects scientific models and theories with events in the world. Science without successful predictions would look a lot like Greek mythology. Pointing out that Zeus is very angry may sound like a nice *explanation* for the tornado in your yard but how to find out if it is true without predictions? More importantly, will your insurance cover angry gods or only angry birds...

> **i** Scientific Predictions
>
> Scientific predictions are statements about data, past events, or the future based on experience or knowledge[4]. This means that prediction is not just about forecasting the future like tomorrow's weather. Predictions can also concern the past. The astronomer Johannes Kepler predicted for instance in

[3]Today, general relativity has strong practical implications as it resulted in making GPS accurate and nuclear energy attainable.

1614 that the Star of Bethlehem was a planetary conjunction appearing in the year 7 BC. The term "prediction" also covers cases in which you already know the correct prediction. For example, if a medical diagnosis tool correctly predicts past cases, it is coherent with the observed data. Nevertheless, the gold standard for validating scientific models and theories is accurate predictions of unknown (future) events or unseen data.

3.4. Prediction serves many purposes

Taking a birds-eye view of science, you find that predictions are everywhere and serve many purposes [26]:

- **Falsifying and confirming hypotheses:** Predictions enable falsification. Predictions that are incompatible with experimental data allow you to reject hypotheses, or weaken scientific theories or scientific models. Conversely, predictions that are compatible with experimental data and future observations boost the confirmation of theories and models.
- **Providing a standard for comparison:** The predictive accuracy of models provides a standard for comparing scientific models that are grounded in reality and independent of how the models work.
- **Checking predictability:** If you can consistently predict a phenomenon based on a set of information, it means that this set contains the relevant information to determine the state of the phenomenon. However, if you cannot, it may be because the available information is insufficient. Knowing what can and cannot be inferred from certain information is immensely important in science [27].

[4] We constrained our definition of scientific prediction to statements based on experience or background knowledge to take an empiricist perspective. Predictions that are based on neither experience nor knowledge belong to the realm of fortune-telling. Comparing different fortune tellers by their predictions, on the other hand, is again a scientific question.

- **Guiding experiments:** Predictions can tell you which aspect of a phenomenon to focus on and run experiments on. If models or theories make interesting or extraordinary predictions for borderline cases, you can check them by running experiments.
- **Enabling practical applications of science:** Science must always be seen as embedded in a larger economic and social process. Predictions can be used for planning, making processes more efficient, and automating them. Predictions are a way in which science pays off societal investments.

> **i** Scientific Hypothesis
>
> Scientific hypotheses are statements about the world that can be tested experimentally and be falsified. These statements can concern predictions or explanations of phenomena. General hypotheses that have been rigorously tested can become scientific theories. The origin of disease in viruses was a hypothesis before it turned into a theory after it had been rigorously tested. In some contexts, entire models are referred to as hypotheses, meaning that the model is the true or best description of the phenomenon within a given class of models.

3.5. Scientific models may interpolate and extrapolate

Weather forecasting and Einstein's relativity theory are great examples of two different modes of prediction:

- **Interpolation** describes the prediction of known data that has been part of the model construction or data highly similar to it.[5] Given a reasonable notion of similarity, interpolat-

[5] In classical math textbooks, *interpolation* usually only describes the prediction of known data. We will use interpolation in the looser sense to denote predictions for highly similar data as it is often used in statistics or social science research.

ing works by a simple principle – treat similar data similarly. Data for which you know the true state forms the basis for predicting similar data for which you don't. For instance, in weather forecasting, you might have a track record of past cloud formations and their corresponding amount of precipitation. Interpolation means to predict the precipitation of new clouds solely based on the precipitation of similar past cloud formations.

- **Extrapolation** is complementary to interpolation. It concerns predicting data for which you have no similar data available. Extrapolation requires more than just treating similar data similarly. For extrapolation, domain-specific background knowledge must be leveraged. The prediction that very heavy stars may collapse into black holes was one consequence of Einstein's theory of relativity and was first drawn by Karl Schwarzschild. "Observations" of black holes didn't exist when Einstein developed his theory. Instead, the prediction was entirely driven by theory, not by reference to similar events in the past.

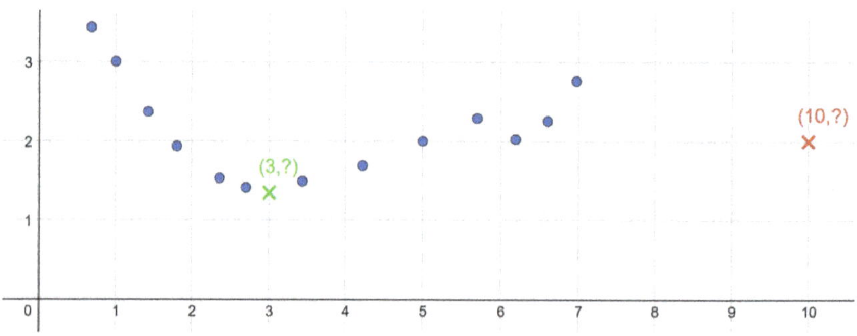

Figure 3.1.: For $x = 3$ (green), y is unknown, but x lies within the range of the data (blue). You have to interpolate. For $x = 10$ (red), y is also unknown, but x lies outside of the range of the data. You have to extrapolate. Figure by [28], CC-BY (https://creativecommons.org/licenses/by/4.0/)

Interpolation and extrapolation play central yet different roles in scientific prediction. Successful interpolation is often taken for granted, as it expresses that the model is consistent with the given data. Extrapolation, or as Popper [29] calls it 'bold hypotheses', on the other hand, takes a more prominent role:

- Successful extrapolation can boost your confidence in the background knowledge that informed the inductive bias you equipped your model with.
- It can spark the belief that the prediction model captured relevant aspects of reality [30].
- Also, extrapolation can guide the experimenting process to try to falsify surprising predictions.

3.6. Machine learning is the ultimate interpolation tool

Supervised machine learning's main strength is the interpolation type of prediction. Supervised machine learning is strongly data-driven rather than relying on background knowledge to predict unknown cases. Extrapolation is therefore not considered a strength of complex machine learning models such as deep neural networks [31], [32], [33]. Some see the current rise of adversarial examples[6] as a direct consequence of such failures of extrapolation [34]. But things are not as simple, for these three reasons:

1. In machine learning, it is possible to use background knowledge to improve extrapolation, as we discuss in Chapter 8.
2. Interpolation and extrapolation can't be separated as cleanly as Figure 3.1 seems to suggest. The separation depends on the similarity of available and new data, and there is no magic cut-off value where a prediction switches from interpolation to extrapolation. A common formal definition of interpolation is the convex hull of the data, however, this would deem almost every prediction in high-dimensional spaces an extrapolation [35]. Given this definition, interpolation would be a

[6]Adversarial examples are slightly modified input data that are predicted incorrectly by the machine learning model.

useless notion outside of low-dimensional and linear prediction problems.

3. With the right notion of similarity in place, every extrapolation can be phrased as an interpolation. Learning a reasonable notion of similarity might be the primary goal of machine learning.

4. Justification to Use Machine Learning

We have demonstrated that prediction is an essential goal in science. Great! But how does machine learning help scientists make predictions? And, what value does it add to existing predictive modeling approaches? What are good reasons for using machine learning in science?

In this chapter, we justify the use of machine learning in science both epistemically and pragmatically.

With the blessing of the Elder Council, tornado prediction became the first machine learning project in Raven Science. Rattle teamed up with the tornado expert Krarah. Krarah was responsible for the data, and Rattle for the training and modeling. Within one month, they had created a model that outperformed all previous tornado prediction models. Eureka! A new paradigm was born.

4.1. Machine Learning has a clear notion of success

In science, the quality of models is often assessed in a lengthy and non-transparent manner. Social aspects such as the reputation of the researchers who developed the model, the opinion of other scientific authorities, or the amount of funding for the research program can play as important a role in model assessment as the predictive capacity of the model [36].

In comparison, machine learning offers a very transparent notion to assess model quality: a good model is one that has a low average prediction error on an unseen test dataset. The idea behind this assessment is that a model with a low empirical error on a test set is expected to make few errors on the task to be solved. This inference from the test to the so-called *generalization error* is even demonstrably correct if the test data is a random and representative data sample of the task.

Furthermore, not only is the goal of modeling very transparent, but the feedback that researchers receive is also quick to obtain and comes in quantified form. In this way, scientists can check their modeling intuitions throughout the modeling process on an almost purely empirical basis and even compare their models with entirely different modeling schools.

This transparency about what makes a good model enables scientists to work towards a common goal. We discuss the concept of generalization in Chapter 7.

4.2. Machine learning adapts the model to the world

Most approaches where mathematical models are fit to data, for example, statistical models or simulations, come with a constrained functional form. Like assuming a linear, exponential, inverse, or additive relationship between variables. Such constraints usually reflect the domain knowledge of the researcher. For example, if you know that a certain fertilizer is less effective in dry climates, you may add an interaction term.

While incorporating domain knowledge sounds smart, it can be dangerous: Your modeling success will depend on the reliability of your domain knowledge and your skill in encoding it. A researcher who only knows linear models will assume linearity in all contexts, whether it is a conscious choice or not. Misspecified models will lead to wrong conclusions and therefore to bad science [37].

In machine learning, the model classes – like neural networks or random forests – are more flexible and powerful. They can, in principle, approximate any function.[1] You don't need to know that there is an interaction between humidity and fertilizer effectiveness to build a good model. The algorithm will learn this interaction from the data. There is a shift in burden, the modeling success rests not on your domain knowledge anymore but on the size and quality of your data. Equipped with enough data, even non-domain experts can find highly predictive models. Indeed, there is a trade-off: with lots of data, domain knowledge is secondary; with little data, on the other hand, domain knowledge is key.

We will discuss the role of domain knowledge and how to incorporate it in machine learning modeling in Chapter 8.

[1] Many hypotheses classes in machine learning are universal approximators of some form; that is, they lie dense within the space of (continuous/step) functions [38], [39]. While this is not a super special property, e.g., also polynomials lie dense in the space of continuous functions, it at least expresses that there is in principle a model that approximates any possible relationship well. This is different compared to classical statistical models that rely on small model classes; even the best model within such a class may be overall a bad model to capture the true relationship between predictors and target.

4.3. Machine learning handles various data structures

The more complex the data structures become, the more difficult it becomes to make assumptions and follow a traditional modeling approach. This becomes evident if we leave the realm of tabular data and consider images, text, and geospatial data. This is where machine learning can shine. Machine learning successfully handles various data types, such as tabular [40], image [41], geospatial [42], sound [43], graphs [44], or text data [45]. There is often lots of data, but it's not always in perfect shape:

- Most data wasn't produced by controlled experiments but comes from the messy world.
- Data is often high-dimensional without indication of what's relevant and what's not.
- Data can come in various forms (text, images, audio, etc.) and is not restricted to tables with clearly interpretable columns.

Unlike approaches such as classic statistical modeling, machine learning can deal with these problems. All you need is data! Well, we exaggerate a bit here, but you get the idea. Dealing with diverse data is not merely practical but also epistemic. If diverse data sources point to similar insights, this strongly supports the validity of these insights [46].

4.4. Machine learning allows you to work on new questions

Machine learning has enabled scientists to leverage data to solve specific problems that they hadn't been able to "solve" before using other approaches:

- **Translating texts:** Machine translation is older than machine learning, but, let's be honest, it was too bad to be useful.

Before machine learning was mature enough and embraced, rule-based systems and statistical machine translation were used.

- **Diagnosis of diseases based on X-ray images:** Before machine learning, techniques like template matching and rule-based algorithms were used. Only machine learning made it possible to make huge progress in this task.
- **Drug discovery:** Traditionally, drug discovery involved a lot of trial and error. Machine learning has made it possible to analyze huge amounts of data and predict how different drugs might interact with targets in the body. In the future, this may greatly accelerate the process of drug discovery.

For such research questions, machine learning is finally offering routes toward studying them systematically, built on empirical data. Paraphrasing Wittgenstein – the limits of my modeling are the limits of my world. This implies that extending the limits of modeling with machine learning extends the realm of phenomena scientists can investigate.

4.5. Don't undervalue machine learning because it is new

Compared to other modeling approaches, machine learning has a short history. Differential equations or linear models, for instance, have existed for hundreds of years. Historically, new modeling approaches like machine learning always had a hard time establishing themselves:

- Einstein's theory of relativity relied on non-Euclidean geometry – an unusual branch of mathematics for these endeavors – which was one of the reasons why he was met with skepticism.
- Probabilistic models were first completely rejected in linguistics due to the dominance of logical methods. Today, they are well-established.

- In econometrics, graphical causal models have a hard time asserting themselves against the prevailing framework of potential outcomes.

The novelty of an approach should neither be taken as an argument in favor nor against using a certain modeling approach. The essential question is: Can the tool help answer the question you have set out to address?

4.6. Further justifications for machine learning

There are further justifications for using machine learning in science:

- **Time efficiency:** Accurate machine learning models can often be built time-efficiently by researchers. There is less preprocessing needed compared to other approaches and large parts of the model selection and training process can be automated. Also, there is great support for machine learning packages in popular programming languages like Python (e.g., scikit-learn, PyTorch, or TensorFlow) or R (e.g., mlr3 or tidymodels).
- **Computational efficiency:** In some situations, machine learning can be computationally cheaper than traditional modeling. Machine learning models can, for example, be used as fast surrogate models for simulations from physics or material science that are computationally very intense [47]. Machine learning weather prediction can run on your laptop, unlike numerical methods that demand the world's biggest computing clusters [48].
- **Basis for theory:** Machine learning may support you with the knowledge needed to build classical theory-based models. They may show which features contain predictive information and how features interact.
- **Effective for operationalized goals:** Machine learning is ideal if the aim can be easily encoded in a single metric. If you

are confident in your metric, machine learning will provide you with the means to optimize for it.

5. Machine Learning and Other Scientific Goals: A Clash

Okay, so the use of machine learning in science can be justified by its predictive power, and prediction is an important scientific goal. All scientists aboard the machine learning ship! Right? While machine learning has been around for years now, why do scientists still have a bad gut feeling about machine learning? In its raw form, supervised machine learning clashes with three other essential goals of science that go beyond prediction: controlling, explaining, and reasoning about phenomena [23], [49].

The great success of tornado prediction increased the number of machine learning supporters. But many voices remained critical. The old veterans, in particular, argued that science is about explaining *why* things happen, not just making predictions. Once again, the Elder Council of Raven Science ended in a heated philosophical debate: Does supervised learning lead to a science we don't understand and can't control?

5.1. Prioritizing prediction over other scientific goals

We believe that the main reason why some scientists are suspicious of machine learning is the exclusive focus on prediction. Bare-bones machine learning readily trades a tiny gain in predictive performance against any other goal scientists have, such as controlling, explaining, or reasoning about phenomena. A supervised machine learning model will take any short-cut to improve predictive performance:

- If given non-causal features that are associated with the outcome, the machine learning algorithm will likely learn a model that relies on them. For example, a machine learning model may rely on the number of sold winter jackets to predict flu cases.
- If the training data are biased, the model will reflect these biases as it helps in predicting the biased data.
- If a complex model has a better predictive performance than an interpretable model, even marginally, the model selection step will spit out the complex model as the winner.

An absolute framing of a question in terms of prediction drowns out other goals. Depending on the context, these goals may have higher importance than prediction.

5.2. Understand action effects to control phenomena

Imagine you study education science. Your data analysis shows a strong association between the academic success of kids and the education level of their parents. In Germany, for example, only 8 of 100 students coming from a working-class family receive a Master's degree, whereas 45 of 100 students with academic parents do so [50]. Indeed, this association allows you to make predictions. Tell me where you come from and I'll tell you where you'll end up. Do you come from a working-class family? The prediction model says: "No Master title for you!"

Focusing on prediction is deeply unsatisfying in this context. Ethically, we as a society do not want this disparity between kids from the working class and kids from academic families – we believe in the equality of opportunity. Instead of predicting academic success, we want to find policy interventions to close the gap. How to encourage more educational climbers? What are the factors that hinder working-class children from pursuing a Master's degree and how to change them? Would an increase in student stipends lead to more students from working-class families?

All these questions are concerned with control. Control is the ability to perform informed actions on a system and estimate the consequential dynamics of the system. For control, we must:

- know the causal factors, know how to act upon them, and also be able to do so
- evaluate the uncertainties in the consequences of actions
- analyze time dynamics and potential long-term equilibria

Let's say the German government decides to increase the financial support for students, known under the beautiful German term Bundesausbildungsförderungsgesetz, short BAföG. They must be sure that a lack of money is one causal factor that prevents students from studying (this is indeed true, see [50]) and have money to increase financial support. The government should be aware of which effect the increase has: is it eaten up by inflation? Can it get more working-class kids to academic degrees? How long will it take for the increase in financial support to be enacted and when will it impact student's choices and how?

Scientific models have always been a way to enable control over a system. They represent the system, allow you to run simulations, and thereby think about the outcomes of actions. However, standard machine learning models are not suitable here. Machine learning models are unable to distinguish causes from effects. Machine learning models neither represent potential actions nor their consequences. Machine learning models only give point estimates rather than uncertainties. Machine learning models often ignore the role of time.

So no control with machine learning? We discuss how at least some of these problems can be alleviated in Chapter 10 and Chapter 12.

5.3. Explain phenomena through counterfactuals

Imagine you study economics. You want to understand how financial crises arise. Your approach is to study historical crises to gain insights into potential future crises. The first thing that comes to your mind is to check the indicators of past crises. Clear signs were plunges in the stock market, increased unemployment rates, or lower gross domestic product (GDP) across countries. All these signs are indicators associated with the crisis. You may even detect them by simply fitting a machine learning model to a dataset containing a broad range of general economic indicators. But is this really what you are after?

Scientists are ultimately interested in *why* financial crises arose rather than *that* they did. Would there have been the great depression in the 1930s without the Wall Street Crash of 1929? How did the Yom Kippur War impact the 1973 oil crisis? What was the role of mortgage-backed security in the financial crisis of 2007-2008?

In all these cases, we are interested in the causes that led to certain outcomes. We want to explain rather than just predict. Explanations point to factors that, if they had been different, would have impacted the explained phenomenon. Think of the oil crisis in 1973. To establish that the Yom Kippur War was a cause, we must establish two conditions: 1) *factual condition*: there was the Yom Kippur War and the oil crisis in 1973; 2) *counterfactual condition*: if there had not been the Yom Kippur War, there would have been no oil crisis. While the first condition can be easily checked empirically, the second condition is tough.[1] No one knows for sure how history turned out to be without the Yom Kippur War. The Organization of Arab Petroleum Exporting Countries (OAPEC) might not have set an embargo or set the embargo later to influence another US policy. Still, scientists can often provide arguments to establish counterfactuals, for example, pointing to other actions of the OAPEC, game-theoretic reasoning, or interviews with OAPEC leaders.

i What establishes a scientific explanation?

Philosophers of science agree on these key ingredients for explanations:

- **Explanandum** (pl. explananda): a statement about the phenomenon to be explained. In the oil-crisis example, the explanandum could be the statement 'There was an oil crisis in 1973.'
- **Explanans** (pl. explanantia): a statement that if true would explain the explanandum. In the oil-crisis example, the explanans could be the statement 'The US supported Israel in the Yom Kippur war.'
- **Explanatory link**: rules or laws describing the mechanism underlying the phenomenon by which the explanans is connected to the explanandum. In the oil-crisis example, the explanatory link could be 'The OAPEC, who stood on the Yom Kippur war with the Arab states, wanted to change the US policy concerning Israel. The OAPEC leader, particularly King Faisal, believed that the oil could be used as a weapon to force the US to change its policy.'

Scientific models are one approach to obtaining scientific explanations. Scientific models represent the explanandum, the factors referenced in the explanans, and, most importantly, the explanatory

[1] The fact that it is impossible to get data to establish the counterfactual condition is called the fundamental problem of causal inference [51].

link (e.g. the causal relationships) to establish a scientific explanation. Unfortunately, standard machine learning models do not serve this function of scientific models. Machine learning models do not distinguish between causal factors and mere associations. Machine learning models do not describe the real-world mechanism that led to the outcome but only give predictions. Machine learning models work well in static environments but are unreliable if the environmental conditions are altered.

So no explanation with machine learning? We discuss how at least some of these problems can be alleviated in Chapter 8, Chapter 10, and Chapter 11.

5.4. Reason about phenomena via representation

Imagine you are a biologist and want to study how the amino-acid sequence influences the protein structure. You know for a fact that the protein structure is largely determined by its amino-acid sequence. Some researchers developed this amazing deep neural network that is very good at predicting the protein structure. Great, problem solved?!

Well, yes and no. You solved one critical aspect of the problem, you can describe the protein structure and this is what ultimately defines the function of the protein. But it seems your solution is not intelligible to us humans – and ultimately, humans are the entities who are conducting science. Why do certain amino-acid sequences lead to specific predicted protein structures? What are the most important subsequences when it comes to the prediction? How does the prediction model align with theoretical biological knowledge about amino-acid sequences and proteins?

Being able to reason about phenomena is important for two reasons:

- **Communication:** Humans must be able to teach and share scientific knowledge in schools, universities, and research papers. This allows us as a species to jointly reason about phenomena and collaborate. Moreover, if scientific knowledge informs public decisions, it must be cognitively accessible to a broad public.
- **Intuition pump:** Intuition drives the questions scientists ask and the experiments they conduct. It allows to draw links to other phenomena or fields, and to traditional scientific models or knowledge. You cannot intuit about phenomena to which you have no cognitive access.

One approach to reason about phenomena is via scientific models. Scientific models provide a language for phenomena. Through variables, functions, and parameters, models deliver a vocabulary (tokens) to talk about phenomenon components, dependencies, and properties respectively [52]. Simplicity and beauty are important in that respect: Simplicity allows humans to communicate effectively and efficiently, and we can emulate phenomenon behavior without running exact simulations; Beauty sparks our curiosity and intuitions, it allows us to learn the scientific language quickly and draw new links.[2]

Standard machine learning models, like the protein structure prediction model from the example above, are unfortunately terrible reasoning tools. Only very few model elements can be interpreted, such as the input (amino-acid sequence) and the output variables (protein structure). Most other model elements do not allow a simple interpretation, think, for example, of the weights or activation functions in the neural net. Similarly, it is hard to assess how well machine learning models align with our background knowledge and also hard to infuse such knowledge. Instead of simplicity and beauty, machine learning models are designed for high complexity and trained solely under the objective of predictive performance.

[2]Interestingly, simplicity and beauty seem to be favored not only by humans but also, at least partially, by nature, as evidenced by the overwhelming success of Ockham's razor in many disciplines when it comes to extrapolation-type predictions.

So no reasoning with machine learning models? We discuss how at least some of these problems can be alleviated in Chapter 8, Chapter 9, and Chapter 14.

6. Bare-Bones Machine Learning is Insufficient

Even though machine learning has great merits when it comes to prediction, it clashes with other scientific goals like control, explanations, and reasoning. This chapter highlights more concretely and practically the insufficiencies of bare-bones machine learning as a scientific methodology.

The veteran scientists' concerns about machine learning were not misplaced. Over time, the tornado system began to act up. False positives and false negatives became more common. Krarah consulted Rattle, but even she was puzzled. Rattle concluded that the problems were not specific to tornado prediction, but systemic to machine learning. They had to be fixed.

6.1. Low test error is not enough

As we highlighted in Chapter 4, machine learning has a transparent notion of what makes a good model: a good model has low error on an unseen test set and in consequence a low generalization error. However, this notion is empty if it is not equipped with an underlying theory of generalization. When can you infer that the model performance generalizes from just knowing the error on a test set? Are there guarantees that certain learning algorithms must yield models with low expected error?

In Chapter 7, we present statistical learning theory as the main contender for a theory of generalization in machine learning.

However, the standard notion of generalization only reflects the predictive capacities of models on data from the same distribution. Extrapolation to completely new scenarios or data from different distributions is not encompassed in the standard notion of generalization. It therefore has limited bite, especially in science, where the goal is to generalize from models to insights about the phenomenon.

In Chapter 7, we therefore also discuss broader conceptions of generalization. Chapter 11 on robustness, moreover, shows how to define generalization under distribution shifts.

6.2. Domain knowledge is overlooked

Science builds on prior knowledge to produce knowledge. Bare-bones machine learning disrupts both ends of this knowledge flow: You expect the model to figure out relations from the data, and get out an opaque prediction model. The focus shifts from epistemology to utility, from science to engineering. You tinker with the model to improve a metric. If machine learning is fully embraced in its bare-bones form, a lot of domain knowledge is left unused, and there is little new knowledge to gain.

However, coherence with background knowledge makes scientific models more valuable. It is smart to incorporate additional information[1]. Compare that to statistical modeling or differential equa-

[1]Usually scientific theories are informed by a lot of historical data; thus, in a certain way incorporating background knowledge increases the data support.

tions: These modeling approaches encourage, even require, the formulation of prior knowledge in terms of distribution assumptions and equations. And you get interpretable estimates back that help you better understand the phenomenon you study.

In Chapter 8, we argue that you can take a domain-knowledge-driven approach with machine learning as well. Even better: Thanks to the focus on predictive performance, you can evaluate your domain knowledge in terms of predictive performance.

6.3. Predictions cannot be easily explained and interpreted

Interpretability enables you to justify models and reason about phenomena. But machine learning models are generally not inherently interpretable, because they may have a complex functional form that is adjusted to data – they are black-boxes. This makes it difficult to understand how the model behaves and what it relies on:

- What were the most influential features?
- Which features did the model learn?
- Why did the model make a certain prediction?

We show in Chapter 9 how to use interpretability techniques to improve models and gain insights from them.

6.4. Predictive performance is at odds with causality

The world we live and act in is a gigantic causal mechanism. Barebones machine learning models however ignore the causal dimension of things. All they care about is making better predictions, and they rely on whatever statistical dependency to pursue this goal. Do you want to know your COVID risk? Alright, the machine learning model needs your shoe size, salary, and the football ranking of your local club.

But scientists want to separate between causes, effects, or spuriously correlated features. Physicians want to know why certain people have COVID and others don't. They want to develop drugs and prescribe treatments that make people healthy. To answer these questions, causality must be taken into account.

We'll show in Chapter 10 how the combination of causal inference and machine learning allows you to find causal dependencies, learn causal models, and estimate causal effects.

6.5. Models lack robustness in deployment

Bare-bones machine learning provides predictive models, but only for a static environment. This means machine learning models are only accurate, if:

- they are applied to data that is similar to the training data, and
- the data structure remains intact during deployment.

In the wild, phenomena are complex. Data changes constantly, external factors enter, measurement devices produce errors, time moves, and observation targets shift. If you want to use machine learning models in your scientific pipeline, you have to make them robust tools under real-world conditions.

We show in Chapter 11 what types of robustness you may be interested in and discuss strategies such as data augmentation that help you robustify your model.

6.6. Predictions come without uncertainty quantification

Proper uncertainty quantification is crucial when high-stakes decisions are made in the real world. However, bare-bones machine learning models only provide point predictions. Some models, such as Bayesian models, come with built-in uncertainty quantification, but limiting the model class to these models may result in a loss in predictive performance. Perhaps the best-performing model is a random forest. If you choose the best-performing model, you might get one without built-in uncertainty quantification. Even when model outputs look like probabilities because they are between zero and one (yes, we're talking to you, softmaxers), they often cannot be interpreted as 'real' probabilities when models aren't well calibrated.

We discuss the philosophy behind uncertainty in machine learning, calibration, and model-agnostic methods for uncertainty quantification in Chapter 12.

6.7. No consensus on standards for reproducibility

There are many standards for how to document code, data, and models for others to reproduce results. Reproducibility is key to:

- being transparent about what you did to achieve high performance,
- allowing others to test your work and gain trust in it,
- reliably building on your results, and
- making your code reusable for potential applications.

Unfortunately, many papers that use machine learning in science do not allow for reproducibility [53]. Often, important information is missing such as on the weight initialization, training epochs, hyperparameters, or random seeds. Also, preprocessing steps might not be listed, and the code is poorly documented or poorly implemented.

We discuss standards for reproducibility of machine learning in science in Chapter 13.

6.8. Missing standards for reporting

In science, it is of central importance to make transparent why a scientific model is suitable and what data it explains. In machine learning, researchers still need to determine what information is scientifically relevant. Predictive performance on the test set is indeed important. But what about the importance of features? Or information about the data collection or the model selection?

We discuss different standards for reporting model results in chapter Chapter 14. We provide a high-level view of how to group these standards and some best practices for publication.

Part II.

Integrating Machine Learning Into Science

In Part I, we argued that machine learning and science could be a great match. However, bare-bones machine learning is an incomplete scientific methodology when it comes to incorporating domain knowledge, gaining causal insight, and modeling uncertainties of predictions.

Part II concerns these missing puzzle pieces and puts them together. The chapters focus on formalizing and specifying the problems. For example, we introduce causal models to formulate causal questions or basic probability theory to talk about uncertainties. We also discuss general solutions like using double machine learning to estimate causal effects or using Rashomon sets to quantify uncertainties. While each solution is presented in isolation, we point out interactions between different topics. Part II consists of the following chapters:

- Chapter 7: Connecting the theory of generalization to machine learning practice.
- Chapter 8: Integrating domain knowledge using suitable inductive biases.
- Chapter 9: Interpreting machine learning models to gain scientific insights.
- Chapter 10: Integrating causal assumptions to draw causal inferences.
- Chapter 11: Analyzing distribution shifts and robustifying machine learning models.
- Chapter 12: Understanding the sources of error and quantifying uncertainties.
- Chapter 13: Avoiding common pitfalls that hinder reproducibility.
- Chapter 14: Providing best practices for model reports.

7. Generalization

Little does Juan know that his chest X-ray was one of the data points for a pneumonia classifier. He presented with a fever and a bad cough at the emergency room, but it was "just" a bad flu. No pneumonia. The chest X-ray that ruled out pneumonia was labeled as "healthy" and later used to train a machine learning model. The pneumonia classifier is not for our imaginary Juan though, because this ER visit was years ago and the case is closed. While the machine learning experts don't care about Juan's images specifically, they care about cases *like* Juan's: Patients coming to the emergency room with symptoms of a lung infection.

That's the promise of **generalization** in machine learning: to learn general rules from specific data and apply them to novel data. To generalize from Juan to many. Without generalization, machine learning would just be an inefficient database. But with generalization, machine learning models become useful prediction machines.

In science, generalizing from specific observations to general principles is a fundamental goal. Scientists usually don't care about specific experiments, surveys, simulations, or studies, but they use them to learn the rules of our world.

This chapter discusses generalization in machine learning and is structured into three parts, each describing generalization with increasing scope.

- **Generalize to predict in theory:** This is the theory of generalization as it is typically understood in machine learning. It concerns key topics from statistical learning theory, such as empirical risk, the IID assumption, and a discussion of the double descent phenomenon and its relationship to under- and overfitting.

- **Generalize to predict in practice:** This section describes a more practical idea of generalization. Rarely does the training setup match the application. To generalize the model to the application requires attention to things like the data-generating process, non-IID scenarios, and distribution shifts.
- **Generalize to understand the phenomenon:** This type of generalization is often implicitly the goal of scientists. It bridges the gap from machine learning theory to scientific applications and discusses data representativeness and the data-generating process.

Nuts are delicious but hard to crack. So the Ravens set out to build a nut quality predictor. Every tenth household had to bring a sample of nuts to Rattle so she could train a machine learning model. The model worked well on the training data, but it was terrible on unseen data. Rattle began to wonder how to ensure that machine learning models generalize.

7.1. Generalize to predict in theory

You want our models to work well on the dataset at hand but also on similar data. One language of similarity is that of statistical distributions. You can think of distributions like a huge bucket that contains infinitely many data points. From this bucket, you can draw data and record it. Think of the bucket that contains X-rays and their corresponding labels. We denote this bucket by the statistical distribution $\mathbb{P}(X, Y)$, where X describes the pixels of the X-ray images and Y the labels.

Equipped with distributions, we can describe more elegantly what our models should optimize. Machine learning models should make as few mistakes as possible in expectation. More technically, they should have minimal *expected loss* $R(\hat{f})$ (sometimes also called *expected risk*):

$$R(\hat{f}) = \mathbb{E}_{X,Y}[L(Y, \hat{f}(X))]$$

This formula describes the expected error the model will make on instances drawn from the distribution bucket $\mathbb{P}(X, Y)$. The "error" for one data point is described by loss function L which quantifies the error between prediction $\hat{f}(x)$ (e.g. pneumonia) and the actual outcome y (e.g. healthy). The problem is that you don't know what the bucket – aka distribution – looks like. You only have a limited amount of data that you recorded. When you have data, you look at the errors the model makes on these data and average over it.

You could use the training data to estimate the expected loss, but using training data makes for a bad estimator of $R(\hat{f})$. The estimated loss would be over-optimistic, meaning too small. If a model overfits the training data ("memorizing" it), the training error can be low even though the model won't work well for new data. It is like preparing students for an exam by giving them the questions and answers beforehand. This means they can simply memorize the answers and you won't get an honest assessment of the student's skills on the subject. The X-ray classifier might work perfectly for Juan and the other training data subjects, but not for new patients. But this has a simple solution: Estimate the expected risk using new data.

$$\hat{R}(\hat{f}) = \sum_{i=1}^{n_{test}} L(y^{(i)}, \hat{f}(x^{(i)}))$$

This formula is also known as test error, out-of-sample error, generalization error, or empirical risk (on the test set).

Slowly but surely, we are piecing together a language to talk about generalization. A model generalizes well when $\hat{R}(\hat{f})$ is low and when the so-called generalization gap is small, which is defined as the following difference [54]:

$$\delta_{gen}(\hat{f}) = R(\hat{f}) - \hat{R}(\hat{f})$$

If the generalization gap is small, the model will perform similarly well for both training and unseen data. [1] Let's explore how the generalization error behaves in different scenarios.

Underfitting and overfitting

Machine learning can feel more like an art than a science, but there is an entire field dedicated to putting all the deep learning magic and mystical random forests on a scientific grounding: statistical learning theory, which provides a view of machine learning from a statistical lens. We explore statistical learning theory to shed light on generalization.

Well-studied concepts are overfitting and its counterpart, underfitting. Underfitting is when the model is not complex enough to model the relation between input and output, so the model will have both a high training and test error but a potentially small generalization gap. Underfitting models are, frankly, bad! Overfitting is when the model function has a bit too much freedom: It fails to capture generalizable rules and instead "memorizes" the training data. That's why overfitting is characterized by a low training error and a high test error and therefore a large generalization gap. Both underfitting and overfitting are undesirable as they both mean a failure to generalize well (measured as low out-of-sample error).

[1] Confusingly, the generalization gap is sometimes referred to as the generalization error.

Going back to the chest X-ray example: Imagine the classification algorithm would be a simple logistic regression classifier based on the average grey scale value of parts of the image. It might work better than random guessing, but wouldn't produce a useful model. A case of underfitting. Overfitting in this same case would look like this: Let's say you use for the chest X-ray a decision tree that is allowed to be grown to full depth. Inputs are the individual pixels and all typical restrictions are lifted, like having a minimum amount of data in each leaf. The tree could grow very deep and separate all training data, meaning each data point gets its leaf node. So the model would work perfectly on the training data. But when used on new data, the decision tree would fail. Figure 7.1 showcases underfitting and overfitting on a simple 1-dimensional case.

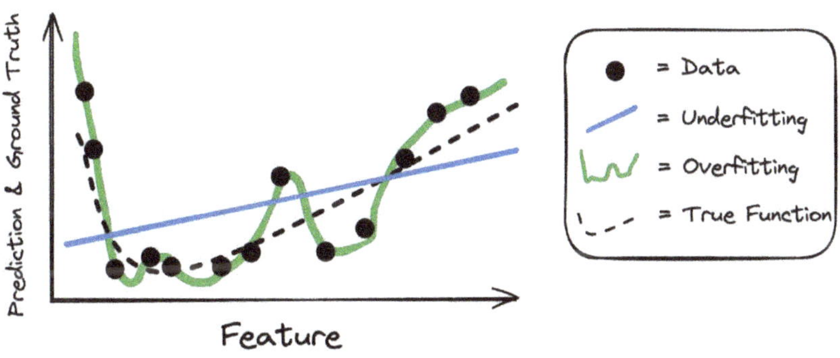

Figure 7.1.: The data (dots) were produced by the true function (dotted line) plus some noise. A well-trained model would approximate the true function well. The linear model (blue line) underfits the true curve, while the too-flexible model (green curve) overfits the data.

Whether a model will underfit or overfit depends on the machine learning algorithm responsible and the complexity of functions it can produce. By picking certain types of model classes and setting their hyperparameters, you can steer the flexibility of the models and therefore the balance between underfitting and overfitting. The typical approach in machine learning is to use fairly flexible models and then regularize them.

Examples of such flexible models are neural networks and decision trees. Theorems show that both neural networks [38], [55] and decision trees can approximate arbitrary continuous functions [56]. These flexible models can then be regularized by specifying certain hyperparameters in modeling such as the learning rate, the architecture, the loss function, or enabling dropout [57].

Underfitting and overfitting don't tell us about the types of errors the models make. This will be covered in Chapter 12 about uncertainty.

Double descent or why deep learning works

We've painted a neat picture of what a perfectly balanced model looks like – models should be flexible enough not to underfit and regularized enough not to overfit. But now with deep learning, the over- and underfitting reasoning doesn't seem to work any longer. Deep neural networks have millions or more parameters and can perfectly fit the training data in infinitely many ways, so you would expect strong overfitting. The thing is – they generalize. It is like in society: the laws of under and overfitting developed for the average John Doe model don't apply to the fancy models rich in parameters. This surprising learning behavior in deep neural networks has been named *double descent* [58]. Double descent describes the out-of-sample error when increasing the ratio between parameters and data. The behavior can be sliced into two components:

1. **Typical under- and overfitting:** The dataset remains fixed and you start with a simple neural network. If you increase the number of parameters in our model and fit it to the data, you observe the typical underfitting and overfitting. This is true until you reach the point where you have as many parameters as you have data points, the so-called *interpolation threshold.* The test error explodes when reaching the interpolation threshold.
2. **Double descent:** But unlike traditional under- and overfitting, the test error decreases if you increase the number of parameters beyond the interpolation threshold. Continuing

to increase the network size, the test error may even become lower than the test error of the "ideal" model in the underfitting/overfitting world (see Figure 7.2).

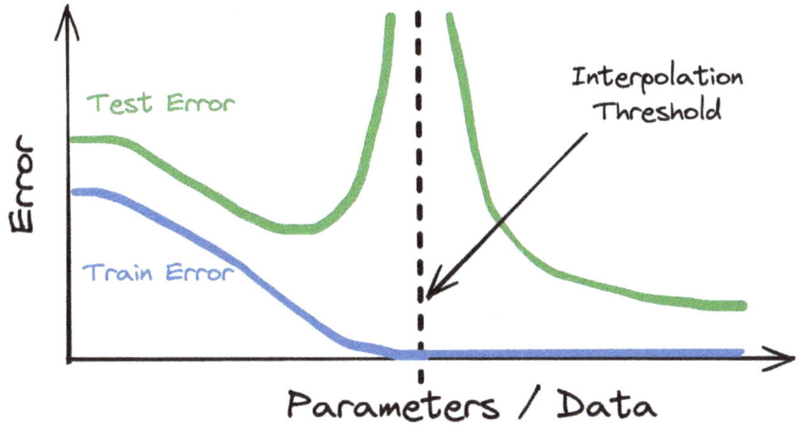

Figure 7.2.: Double Descent. Image inspired by [59]

Double descent is not exclusive to deep neural networks but also happens for simple linear models [60], random forests, and decision trees, as suggested by [58], possibly due to a shared inductive bias [61]. Double descent undermined the theory of underfitting versus overfitting. But under- and overfitting are still useful concepts. It is like with Newton's theory of gravity when Einstein's relativity came along: Underfitting and overfitting provide an accurate picture of things below the interpolation threshold, but beyond this threshold the classical picture becomes invalid.

Double descent describes the what but not the why. We still have no definitive answers as to why overparameterization works so well, but there are theories:

- The lottery ticket hypothesis [62] says that there are subnetworks in certain trained neural networks that have similar performance to the overall network. Training a large network is like having multiple lottery tickets (aka subnetworks) and one will win.

- Benign overfitting [63]: Many low-variance directions in parameter space are required to achieve highly performing models. This is achieved through overparameterization and makes for "benign overfitting".
- Implicit regularization [64]: Optimization algorithms such as stochastic gradient descent implicitly regularize the model. It was shown that stochastic gradient descent actually optimizes not only the loss but effectively the loss plus an implicit minimizer.

We barely scratched the surface of statistical learning theory, and there are many more topics to explore:

- Quantifying the complexity of models (like VC dimensions).
- Learning guarantees for kernel methods like support vector machines.
- Studying consistency and convergence rates of learners.
- Providing bounds for the empirical risk.

7.2. Generalize to predict in practice

So far we've talked about generalization from a theoretical viewpoint that, in practice, is too narrow. Because in practice, you only have access to data but not to the underlying distributions. Data is messy, noisy, and cannot perfectly be trusted.

Generalization through splitting data

How do you obtain models that generalize while being data-efficient? The answer: data splitting! Let's explore this with an example: Rajpurkar et al. [65] built a chest X-ray image classifier to detect pneumonia. To ensure that the classifier generalizes to new data, they split the data into training data (93.6% of the data), validation data (6%) to control the learning rate, and test data (0.4%) to evaluate the final model. If they had used 100% of the data for training the model, they would run into two problems:

1) The model might perform badly since it is unclear how many epochs to train it, and 2) the modelers would have no idea about the performance of the model, except for an overly optimistic estimate on training data.

But if you split the data, train a model on one part, and evaluate the model on the remaining part, you can get an honest estimate of the out-of-sample error. Great, problem solved?! Careful, while their approach gets them an unbiased estimate of the test error, the estimate possibly has a large variance. With only 420 images in the test set, 10 difficult cases that ended up in the test set by chance can spoil your performance estimate. One strategy to lower the variance is to split the data more often. For example with cross-validation: Split the data, for example, into 5 parts, combine 4 parts for training (and validation), and the remaining 1 part for testing. Repeat this setup 5 times so each part is once used as test data. Average the 5 estimates of the out-of-sample error and, voila, you have a more stable estimate (visualized in Figure 7.3).

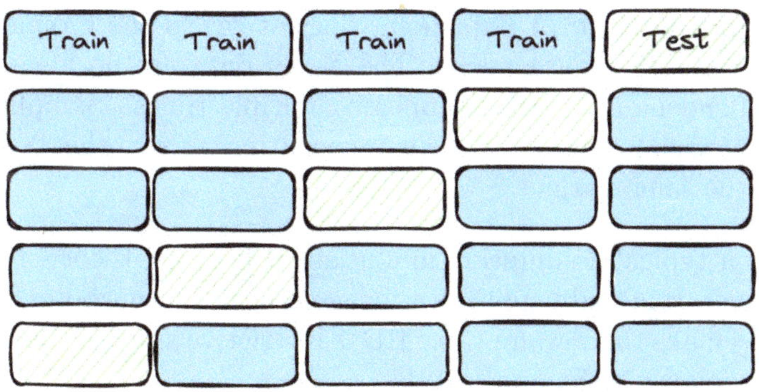

Figure 7.3.: 5-fold cross-validation

But there's another problem. In each CV-loop, you split the data once into training and validation data. The validation data in [65] was used for adapting the learning rate, but you could also use it for hyperparameter tuning and model selection. A single split can lead to a similar problem as before: too much variance in the performance estimate. So you might want to have another cross-validation inside the outer cross-validation. This so-called nested cross-validation quickly blows up the number of models you have to train, but it is a more efficient use of your data. This quickly went from splitting the data into two parts (training and testing) to splitting the data 100 times (10-fold cross-validation within 10-fold cross-validation). Data splitting is at the heart of generalization.

The tricky IID assumption

Statistical theory and data splitting practices rest on a crucial assumption: data are IID, which stands for "independent and identically distributed" and means that each data point is a random sample.

- Identically distributed: All the data points are from the same distribution and don't change over time. If you had one set of X-ray data for model training from a children's hospital but the model application from an adult hospital, they are not identically distributed.
- Independent: A data point doesn't reveal the ground truth of another data point. The X-ray data are no longer independent if a patient appears multiple times. Sampling one X-ray of a patient reveals information about other X-rays of the same patient.

IID is a typical assumption in statistical learning theory, but also when you randomly split data for generalization purposes you implicitly make this assumption. IID is restrictive and real-world data often violates it. Some examples:

- Store sales over time are not IID.
- Patient visits with possibly multiple visits per patient are not IID
- Satellite images of neighboring locations are not IID.

An earlier version of the paper by Rajpurkar et al. [65] ran into this non-IID problem: They split the data randomly, but for some patients, there were multiple X-ray images in the data. This led to data leakage: The model had an easier job since the model was able to overfit patient characteristics (e.g. scars in the X-ray image) and that would help classify the "unseen" data. As a kid, our imaginary Juan fell from a tree and broke his rips. This past injury is still visible in chest X-ray images and uniquely identifies Juan. If Juan went multiple times to the emergency room, his images might end up in both training and testing, and the model may overfit on the scans.

Rajpurkar et al. [65] fixed this problem by ensuring that a patient's data can only be in training or testing, but not both. If IID is violated, generalization can break down in parts – unless we account for it. The IID assumption also helps us in estimating the test error: If the data are IID, we can estimate the generalization error in an unbiased way because of the law of large numbers.

The real world is messy

When COVID hit, many machine learning research labs dropped their projects to work on COVID detectors, many of them from X-ray images. Partially understandable, but in hindsight, a waste of effort. Sounds harsh, but Wynants et al. [66] did a systematic review of 232 prediction models for COVID and found that only 2 (!) were promising. The remaining 230 had various problems, like non-representative selections of control patients, excluding patients with no event, risk of overfitting, unclear reporting, and lack of descriptions of the target population and care setting.

If you want a functional COVID-19 X-ray classifier, you should be as close as possible to the data-generating process of a potential application. For instance, getting data directly from an ER where radiologists label the images with the diagnoses. This would generate a dataset that reflects a typical distribution of cases. However, the data that many machine learning labs used were quite different. So different that the research models and results are unusable. As the pandemic progressed, more and more X-rays of COVID-infected lungs were posted online in repositories. Often without metadata like missing demographics of the patient, without any verification process, and little documentation. But that's not the worst part of COVID classifiers. For classification tasks, you also need negative examples, such as images of healthy lungs or from patients with, for example, pneumonia. These negative images were cobbled together from many pre-pandemic datasets. A red flag: Negative and positive X-ray data come from very different data-generating processes. Should a deep learning model find any hints or shortcuts that identify the data source, then it doesn't have to detect COVID at all. But even that isn't the worst yet. The worst is how the non-COVID dataset was assembled. Roberts et al. [67] looked more deeply into the most commonly used datasets and found the following fouls:

- The X-ray image datasets were put together from multiple other image datasets.
- One of these datasets was from children (only non-COVID).
- Some datasets were included more than once, leading to duplicated images, introducing non-IID problems and data leakage.
- For some of the datasets it is intransparent how they were collected
- Other datasets were collected through "open calls" to other researchers to submit data without further verification.

These things should all raise red flags. It is like Frankenstein was employed to create a dataset. A data-generating process that deviates strongly from any application we can think of. A model trained on Frankenstein's data can learn all matters of shortcuts and none will generalize to a meaningful application:

- Identify children's lungs: If the model can identify that the image was from a child, it can safely predict "not COVID".
- Identify the year: If the model can identify the year through explicit or implicit markers (like the type of machine) it can safely label "not COVID" for older images.
- Identify the dataset: Any characteristics that images from the same dataset share can be used to make the prediction task easier. It is enough when a dataset is processed differently (e.g. greyscaling) or comes from a different X-ray machine.

- Duplicates: Some images might have ended up both in training and test data, making the model seem to work better than it does.

Even if you find a model that perfectly predicts identically distributed data, the models can't be used. No application comes with a data distribution anywhere identical to this mess.

In general, to generalize from training to application, you want the data-generating process considered in training to be as similar as possible to the one during deployment. It is difficult. The world is even messier than what we described here and there are many more challenges to generalization in practice:

- **Distribution Shifts:** Imagine someone building a pneumonia classifier before COVID-19. COVID introduced a new type of pneumonia and due to lockdowns and social distancing, other types of pneumonia occurred less frequently. A massive distribution shift may worsen the performance of existing models. Distribution shifts are discussed in Chapter 11.
- **Non-causal models:** The more a model relies on associations but not causes, the worse it might generalize. See Chapter 10.
- **Using an unsuitable evaluation metric:** While this may not show up in a low test error, picking a metric that doesn't reflect the application task well will result in a model that transfers poorly to the real-world setting.

7.3. Generalization to understand a phenomenon

Generalization to predict other data is one thing, but especially in science you often want to generalize insights from the model to the phenomenon you are studying. In more statistical terms this is about generalizing from a data sample to a larger population.

Generalization of insights may even come in innocent ways that we don't immediately recognize. For example, Rajpurkar et al. [65]

claimed that their X-ray classifier performs on par with radiologists, even outperforming them on certain metrics. We could say they only refer to the test data and leave it at that. However, nobody is interested in the test data, but in the population they represent. Like a sample of X-rays taken typically in the emergency room. Unfortunately, the paper doesn't define the population, which is typical for machine learning papers.

When a researcher studies a phenomenon using machine learning and interpretability, such as the effect of fertilizers on almond yield (like [9]), they are also generalizing. They generalize, explicitly or implicitly, from their model and data to a larger context. Quoting from the abstract of [9]:

> We also identified several key determinants of yield based on the modeling results. Almond yield increased dramatically with the orchard age until about 7 years old in general, and the higher long-term mean maximum temperature during April–June enhanced the yield in the southern orchards, while a larger amount of precipitation in March reduced the yield, especially in northern orchards.

The larger context depends on what the data *represents*. In the case of the fertilizer study, this might be all the 6,000 [68] orchards in California. Or maybe it is just the ones in Central Valley? It depends on how representative the dataset is. The word representativeness or especially representative data is overloaded and people use it differently in machine learning [69] and science [70]. In the broadest sense, "representativeness concerns the ability of one thing to stand for another—a sample for a population, an instance for a category" [70]. In machine learning some claim representativeness without argument, some claim non-representativeness because of selection biases, some mean that the sample is a random sample from the distribution, and some claim coverage in the sense that all relevant groups are covered (maybe not in the same frequency as target population though), some speak of it as prototypes and archetypes. But for science and especially for the goal of inference – to draw conclusions about the real world – you need the data to represent the target population, in the sense of the training data being a random sample from the population.

In an ideal world, you start with your research question and define the population. Then you draw a perfectly representative sample because you can just randomly sample from the population, as easy as buying fresh bread in Germany. But that's often far from reality.

The other way would be to start with a dataset, argue which population it represents, and extend insights to this population. And sometimes it is a mixture of bottom-up and top-down approaches. Zhang et al. [9], for example, describes that they collected data from the 8 major growers that make up 185 orchards in the Central Valley of California. Some in the northern, some in the central, and some in the southern region. However, they do not discuss whether their sample of orchards is representative, so it is unclear what to make of the results.

Proving that your data is representative is difficult to impossible. If you know the population statistics, you can at least compare summary statistics between the training set and the population. As always, it is easier to disprove something: finding a single counter-argument is enough. For representativeness, the counter-arguments are called "selection biases". Selection biases are like forces in your collection process that either exclude or at least undersample some groups or over-emphasize others. Selection bias is a good angle to view the collection process. If you have identified a selection bias, you can discuss its severity and maybe even counter it by weighting your samples. Some examples of selection biases include:

- Survivorship bias: The sample only includes "survivors" or those whose objects/subjects passed a selection process.
- Non-response bias: Human respondents can differ in meaningful ways from non-respondents.
- Exclusion bias: Some exclusion mechanism (e.g., due to missing data) biases the sample.

7.4. No free lunch in generalization

We structured this chapter along three types of generalization: to predict in theory, to predict in practice, and to understand a phenomenon. One of the most well-known theoretical results – the so-called no-free lunch theorems – has taught us that generalization never comes for free [71]. All versions of the theorems highlight the following: You will never have an ultimate learning algorithm that always spits out the best possible prediction model [72]. You must take an *inductive leap* to generalize from a data sample to anything beyond itself. Like making context-specific assumptions (e.g. smoothness or IID) [73]. There ain't no such thing as a free lunch, if you want to eat different meals, you need different cooking recipes.

And, unfortunately, there is no free dessert either. Even if you have a model that generalizes well to identically distributed data, you have to "pay" for any further generalization. When it comes to generalization from training to application or from sample to population, you need to make even more assumptions and put in extra effort. And sometimes you might not achieve them after all. Generalization is never free.

The cost of generalization comes up in other chapters as well:

- When interpreting the model for the goal of understanding the phenomenon of interest, you make assumptions about representativeness for example (see also Chapter 9)
- For causal inference, you make assumptions about the causal structures in the world (Chapter 10)
- Robustness is about guarding your models against distribution shifts (Chapter 11)

8. Domain Knowledge

Machine learning algorithms learn models from data without needing much input. At least compared to other modeling approaches. But what about all the domain knowledge that you already have? Having models learn from data seems like the opposite of using domain knowledge – you let an algorithm identify the relevant patterns and differentiate between signal and noise.

> **i** Domain knowledge
>
> Domain knowledge refers to facts that have been established in a field.

Can you use domain knowledge to guide the model? This chapter shows that you can, from the standard steps of translating the problem into a prediction task to more creative means such as designing custom evaluation metrics. The chapter focuses on ways to infuse domain knowledge directly into the model. Less well-known is that you can extract domain knowledge from the model: Domain knowledge constrains the model and affects its performance, which gives us information on how valuable the domain knowledge is for this prediction task. This makes machine learning a two-way street between domain knowledge and predictive performance.

Krarah and Rattle brainstormed about the problems with the tornado prediction system. Rattle had tunnel vision and looked only for reasons in her theory of generalization. Krarah shouted in frustration: "You're not even using the knowledge about tornadoes, which we've accumulated over centuries. No wonder it doesn't work!" This call from reality was sorely needed for Rattle. But how could she integrate domain knowledge into machine learning?

8.1. Translate the question into a prediction task

The Ravens were wrong. Machine learning models are typically infused with lots of domain knowledge already, even when it may seem otherwise at first. Most domain knowledge comes from "translating" the scientific question into the prediction task. Predicting tornadoes from radar data and not from the number of people reporting knee aches might seem common sense. But as soon as you get into the details of designing the prediction task, you typically rely on deeper domain knowledge. Designing the prediction task might not always feel like infusing domain knowledge, especially when the prediction task is long established, like in weather forecasting. But for scientific problems that aren't typically set up as prediction tasks, the translation process allows for a lot of creativity, which should be guided by domain knowledge.

A big part of the translation is creating features and targets. Whether you predict intelligence from the income of the parents or chocolate consumption [74] makes for very different models. But if you are a typical reader of this book, you already know these things. Time to get to the juicy and often overlooked means to infuse domain knowledge.

8.2. Constrain the model

Machine learning is praised for the flexibility to learn any function, but sometimes you might want the opposite and constrain the model. Model constraints are a direct way to incorporate domain knowledge into the learning process. Some examples of constraints that can be put on the relation between features and the prediction:

- **Monotonicity:** The model's predictions have to increase or decrease monotonically with increasing values of the feature in question. Example: Ensuring credit risk scores decrease monotonically with income.
- **Linearity:** Restrict the model to learn a linear relation between prediction and feature(s). Example: Predicting rent, where the price is linearly dependent on the size of the house.
- **Sparsity:** Restrict the number of features to be used for the prediction. Example: In genomic studies, identifying a small subset of genes responsible for a particular disease.
- **Smoothness:** Restrict the predictions to smoothly change when the input changes. Example: Predicting temperature as a function of altitude, where abrupt changes are not expected.
- **Cyclicity:** A smoothness constraint that ensures that the predictions for the low and high end of a feature "meet". Example: Predicting electricity demand based on the time of the day, considering the cyclical nature of a day, including a smooth transition from 23:59 to 00:00.
- **Range constraints:** Restricting the domain of the predictions. For example, pH levels of a solution can only be between 0 and 14.

How do you infuse these constraints into your models? How do you force a support vector machine to model one of the features linearly, but not all the other features? How do you make sure a random forest becomes sparse? Sometimes you can work with model-agnostic methods, such as feature pre-processing:

- To obtain sparse models, use a feature selection step as pre-processing, followed by your usual machine learning pipeline.
- To model a feature cyclically, transform the feature into two features with sin and cos transformations [75].

However, for other constraints, it is necessary to limit the model classes to ones that can handle the constraint. For instance, if you want your models to have monotonicity constraints for some features, you should only use model classes such as:

- Linear regression models.
- XGBoost, a library for a gradient boosting algorithm, which allows setting a monotonicity constraint.
- Neural networks with monotonicity constraints [76].

The more constraints you add, the smaller the pool of eligible model classes becomes. And the more specific the constraints, the more you depart from well-trodden paths, and you'll find yourself using untested code from obscure machine learning publications instead of scikit-learn and other well-established machine learning libraries.

If you find yourself wanting many model constraints that can't be enforced in a model-agnostic way (e.g. feature engineering), you might want to look at frameworks and model families that are constraint-friendly, but also well-established:

- **Model-based boosting** [77]**:** A gradient-boosting-based approach allowing you to pick constraints for each of the features.
- **Deep neural networks:** You can design many constraints through choices of the neural network architecture, loss function, and how the network is trained.

Regardless of whether you add constraints through feature engineering or restrict the model classes, ensure to also fit an unconstrained model. As stated in the beginning, domain knowledge doesn't only flow in the direction of the model, but model performance may tell you something about the predictiveness of that domain knowledge. By comparing the performances of constrained and unconstrained models, you can evaluate the constraint. A decrease in predictive performance when adding constraints might suggest that the domain knowledge behind it isn't as robust as initially thought (unless you can argue otherwise).

8.3. Design your performance metric

There's a temptation to measure model performance with an off-the-shelf metric such as the mean squared error or the F1 metric. An off-the-shelf metric can be fine, but you might be better off designing a custom metric. The performance metric is an excellent, but often overlooked, way to incorporate domain knowledge into the model.

Poor modeling choices, lack of hyperparameter tuning, or bad features result in low predictive performance. That means the performance evaluation acts as a warning system. But if you've put an unsuitable warning system in place, there is no "meta-warning system" that will alert you. Except for reality, because at some point you realize that the model doesn't fulfill its ulterior purpose. The lack of feedback makes it important to pick or design the right evaluation metric. Every metric makes judgments about how bad certain prediction errors are. For example, the mean squared error $\frac{1}{n}\sum_{i=1}^{n}(y^{(i)} - \hat{f}(x^{(i)}))^2$ comes with the following judgments:

- **Symmetry:** Missing the outcome by -1 is as problematic as being off by +1.
- **Equality:** Each data point's prediction is equally important.
- **Non-linearity:** Being off by 2 is four times as bad as being off by 1.

Designing these "judgments" is an opportunity for leveraging domain knowledge. You might weigh certain errors higher, weigh a subset of your data higher, or anticipate distribution shifts.

Let's illustrate this with an example. Goschenhofer et al. [78] developed a wearable-based Parkinson's severity monitor that predicted disease severity (y) from severe slowing of movements (-4) to okay (0) to severe excessive movements (+4). Together with medical doctors, the authors identified three requirements for the performance metric: non-linearity, asymmetry, and lack of translation invariance. The metric should penalize larger errors more heavily. Particularly, errors with a sign change should be more costly. If the actual outcome is -1, a prediction of +1 is worse than -3 because getting the direction wrong is more harmful to the patient than an overestimation. Furthermore, diagnostic errors for patients with severe symptoms should have greater weight than errors for patients with mild symptoms. This is what the new performance metric looks like:

$$\mathbb{P}(Y, X, \hat{f}) = \frac{1}{n} \sum_{i=1}^{n} \left[\frac{y^{(i)}}{4} \alpha + sign\left(y^{(i)} - \hat{f}(x^{(i)})\right) \right] \left(\hat{f}(x^{(i)}) - y^{(i)}\right)^2$$

At its core, it is the squared loss between model prediction and ground truth label but multiplied with a factor controlling over- and underestimation, steered by the parameter $\alpha \in [-1, 1]$, and a sign error penalty. This controls the symmetry (negative α penalizes underestimation, positive α penalizes overestimation, and 0 means symmetric loss). Based on feedback from medical experts, the authors chose $\alpha = 0.25$. By multiplying α with the true label, the true label controls the direction of the asymmetric penalization. For example, classifying a true y=-1 as -3 "costs" 4, but if you classify it as +1, which is also 2 units apart, it costs +5. That's the custom "sign error" penalty of the metric. Figure 8.1 visualizes the loss function.

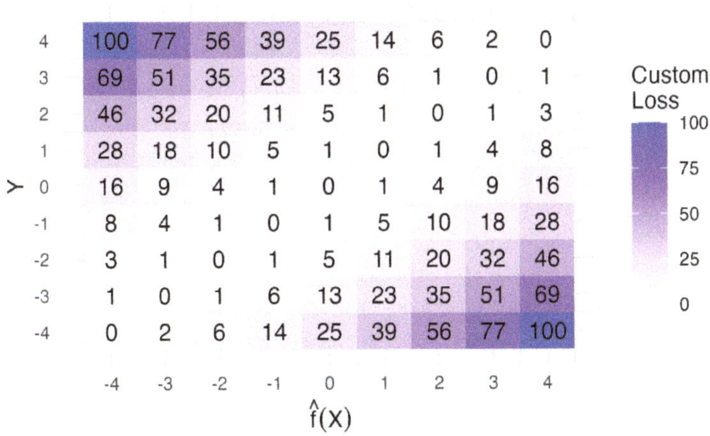

Figure 8.1.: Custom Loss Function

8.4. Align loss and evaluation metric

Ideally, the metric you choose can be used as a loss function, the function that is directly optimized during the training. However, this might not always be possible:

- The performance metric might not be differentiable, but, for example, training neural networks requires a loss function with a gradient.
- Some model classes have a "built-in" loss function, like the linear regression model which optimizes the squared loss by default, or decision trees that optimize the Gini metric.
- The performance metric might be difficult to optimize because it is non-convex or infeasible to compute.
- The performance metric might only work on an aggregation of your data (e.g. F1 score), but the model requires an instance-wise computable loss function.

Here are some practical tips for working with loss functions and performance metrics:

- If possible, design the performance metric so that it is convex and has a gradient. This way, you can use it directly as a loss function, at least for some model classes such as neural networks.

- If you can't develop the perfect loss function, at least align the loss function with the performance metric: Do they at least have the same optimum? Do they model the same (dis)similarities? Do they both have the same scaling? Think linear versus squared.
- Ensure that hyperparameter tuning and model selections are based on the performance metric and not the loss function.
- If you have different options for loss functions, you can treat the loss as a hyperparameter.
- Even if the loss function optimized by some model class seems like a mismatch, train the model anyway and let the model selection process decide. In the case of a huge mismatch between loss and performance metrics, it should show up in the performance evaluation.

8.5. Make it physical

The loss function is also a means to model physics. Many physical processes are guided by partial differential equations (PDEs). These equations are relevant in predicting the weather, simulating the climate, understanding how aircraft fly, and understanding fluid dynamics. However, simulating PDE-based systems can be computationally costly and slow. Wouldn't it be great to use machine learning instead?

At first glance, these equations seem the opposite of what machine learning entails: They represent explicit physical laws, whereas neural networks are often seen as a mess of matrix multiplications. With physics-informed neural networks (PINNs) [79], however, you can make a neural network learn physical laws/PDEs as well, making the neural network an emulator of a physical process. This leverages the property of neural networks being universal function approximators [38]. The trick lies in using a loss function that minimizes a two-part MSE:

$$MSE = MSE_u + MSE_{\hat{f}},$$

where MSE_u corresponds to the initial and boundary conditions of the system, and $MSE_{\hat{f}}$ is the squared error between model output and the output of the original differential equations, based on collocation points, which are points in the domain where the PDE is enforced. Instead of purely emulating the process, physics-informed neural networks can also be used to discover the parameterization of PDEs from noisy measurements of the system [80].

Alternatively, instead of encoding PDEs via the loss function during training, you can also encode physics constraints directly into neural networks, resulting in what some refer to as Physics-encoded Neural Networks (PeNNs).

8.6. Considering various evaluation metrics

Supervised machine learning has a drawback: the evaluation is typically based on just one dimension, the performance metric. But life is never that simple! You might have multiple metrics that you care about. Fortunately, it is possible to train models that optimize different metrics, but it is not as simple as just having one metric.

Imagine you want to predict tornadoes in the next hour. We could have accuracy or F1 score as one metric, but there's more to life than just being right all the time. You might also want to consider the complexity of the model, like how sparse it is, or even the average time it takes to make a prediction (because nobody likes waiting, especially if there's a tornado looming).

There are two options to accomplish this:

- Combine both metrics into one. This only works when you can quantify your desired trade-off between the two metrics.
- Use multi-objective optimization for hyperparameter tuning and model selection.

Here are a couple of metrics that you can optimize for, partially based on [81]:

- **Predictive Performance**: Can be quantified using measures such as F1-Score, AUC, or mean absolute error.
- **Interpretability** [82]: Metrics such as sparsity, interaction strength, and complexity of the main effects.
- **Fairness**: Measured as disparate impact, equalized odds, or calibration.
- **Robustness**: Measured via performance under perturbations, presence of adversarial examples, and under distribution shifts.
- **Alignment with physics-based models** [83]
- **Inference time**
- **Memory requirements**

Let's say you want to predict a disease outcome using gene expression data. From domain knowledge, you know that only a few genes might be relevant. But if you only optimize for predictive performance, the model might pick up more genes than necessary if it doesn't hurt the performance bottom line. By also optimizing for sparsity in this example, you can find a model that satisfies both performance and sparsity[1]. We've already discussed constraints in a previous section, but multi-objective optimization offers a slightly different approach to constraints. Optimizing for multiple objectives produces multiple "optimal" models. Each model is optimal regarding a different trade-off between the objectives, also known as "Pareto-efficient" models. Having multiple models can be both a bug and a feature. It is a bug because you typically want just one model. It is a feature because the set of Pareto efficient models allows you to choose the trade-off between the metrics that make sense based on domain knowledge. And that's easier than having to decide beforehand how to balance both performance metrics.

Again, we can observe the two-way street: Whether you collapse

[1]Multi-objective optimization is not the only method to introduce sparsity into your model. Generally, regularization is employed, implying that only models with inherent sparsity, such as LASSO (a sparse linear model), are considered.

the objective into one function or embrace the multiplicity of models with different trade-offs – multi-objective optimization will enable you to explore different trade-offs and therefore evaluate domain knowledge.

8.7. Use the right inductive biases and learn from them

> **i** Inductive bias
>
> The prior assumptions a machine learning algorithm imposes on the model to help generalize to unseen data.

Imagine a machine learning algorithm that produces "models" that only memorize training data. To "predict", this model would have to check whether the new data point is in the training data. If so, the model can return the outcome for the training data point. If not, it can't make a prediction. But what if a training data point is almost identical to the new one? Could you just use the same prediction? The model could then just return the prediction of the nearest data point. Or, to make it robust, the average prediction of multiple close data points from the training data. That's the idea behind K-nearest neighbors.

But you could also have a model that makes different assumptions about how to generalize to new data, see Figure 8.2. If you get a new data point and one of the features is slightly different from one of the training data points, you could interpolate linearly between the neighboring data points for this new data point. If you do that for all features and training data points, then you end up with a linear regression model.

Or you could use another strategy: Divide your data into partitions which are defined by binary decisions in the features. The partitions are chosen so that the training data points within have a similar outcome. This comes with a beautiful generalization strategy: For each new data point, you just have to check in which data partition

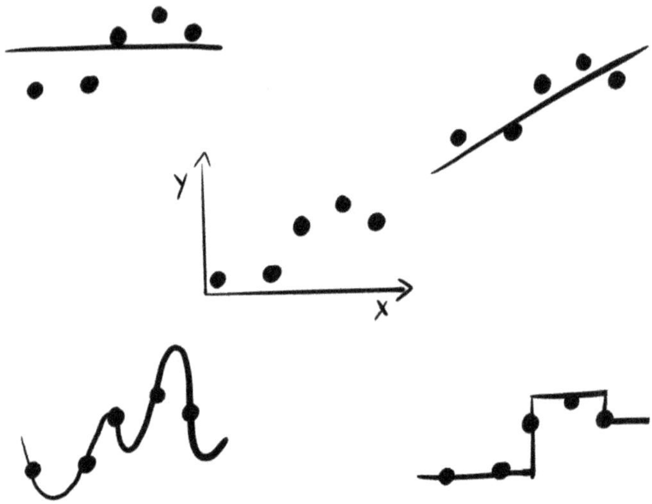

Figure 8.2.: Same data, different inductive biases.

it falls, based on the feature values, and then take the average outcome of all the training data points as a prediction. This is the strategy that decision trees use.

All these examples show that each model class, like trees, k-nearest-neighbors, and linear models, comes with different inductive biases. The inductive bias is an instruction on how to generalize the relation between inputs and outputs to unseen data.

Try out models with different inductive biases. If a certain inductive bias seems to stand out, study how it relates to the studied phenomenon. Can you learn about your problem at hand?

A way to understand inductive biases better is by using tools from interpretable machine learning (see Chapter 9). Interpretable machine learning can also be used to extract insights from the models. Domain knowledge often comes in the form of causal knowledge (see Chapter 10).

9. Interpretability

What's your biggest concern about using machine learning for science? A survey [84] asked 1600 scientists this question. "Leads to more reliance on pattern recognition without understanding" was the top concern, which is well-founded: Supervised machine learning is foremost about prediction, not understanding (see Chapter 2).

To solve that problem, interpretable machine learning was invented. Interpretable machine learning (or explainable AI[1]) offers a wide range of solutions to tackle the lack of understanding. Interpretability, in the widest sense, is about making the model understandable to humans [87]. There are plenty of tools for this task, ranging from using decision rules to applying game theory (Shapley values) and analyzing neurons in a neural network. But before we talk solutions let's talk about the problem first. Interpretability isn't a goal in itself. Interpretability is a tool that helps you achieve your actual goals. Goals that a prediction-focused approach alone can't satisfy.

[1]What do interpretability and explainability exactly mean? Even researchers in this field can't decide on a definition [85]. From an application-oriented perspective, it is most useful to treat these terms interchangeably. Under these keywords, you find approaches that allow you to extract information from the model about how it makes predictions. We use the definition from [86]: Interpretability is about mapping an abstract concept from the models to an understandable form. Explainability is a stronger term that requires interpretability and additional context.

The integration of domain knowledge eliminated many childhood diseases of machine learning. The Ravens began building larger and larger models to study increasingly complex phenomena. Krarah approached Rattle with a question: "The models must have learned so many interesting relationships, is there any way to extract this knowledge?" Rattle nodded and winked at Krarah. Recently, Raven Krähstof wrote a book about interpretable machine learning.

9.1. Goals of interpretation

Imagine you model the yield of almond orchards. The prediction model predicts almond yield in a given year based on precipitation, fertilizer use, prior yield, and so on. You are satisfied with the predictive performance, but you have this nagging feeling of incompleteness that no benchmark can fill. And when you find a task can't be solved by performance alone, you might need interpretability [88]. For example, you might be interested in the effect of fertilizer on almond yield.

We roughly distinguish three interpretability goals (inspired by [89]):

- **Discover**: The model may have learned something interesting about the studied phenomenon. In science, we can further distinguish between confirmatory (aka inference) and exploratory types of discovery. In the case of almond yield, the goal might be to study the effect of fertilizer.

- **Improve**: Interpretation of a model can help debug and improve the model. This can lead to better performance and higher robustness (see also Chapter 11). Studying feature importance you might find out that the prior yield variable is suspiciously important and detect that a coding error introduced a data leakage.
- **Justify**: Interpretation helps to justify a prediction, or also the model itself. Justification ranges from justifying a prediction towards an end-user touching upon ethics and fairness, to more formal model audits, but also building trust through, e.g. showing coherence with physics. For example, you are trying to convince an agency to adapt your model, but first, you have to convince them that your model aligns with common agricultural knowledge.

The goals interact with each other in the life cycle of a machine learning project. For example, to discover knowledge, your model should show good performance. However, building a performative model is an iterative process in which interpretability can immensely help by monitoring the importance of features.

But how do you achieve interpretability for complex machine learning models? To oversimplify, the field of interpretable machine learning knows two paths: interpretability-by-design and post-hoc interpretability. Interpretability-by-design is about only using interpretable, aka simple models. Post-hoc interpretation is the attempt to interpret potentially complex models after they were trained.

9.2. Interpretability-by-design

Interpretability by design is the status quo in many research fields. For example, quantitative medical research often relies on classical statistical models, with logistic regression being a commonly used model. These statistical models are considered inherently interpretable because they often relate the features to the outcome through a linearly weighted sum in one way or another. Statistical models are also dominant in the social sciences and many other fields. The use of a linear regression model to predict almond yield is not unheard of [90]. However, models with very different motivations can also be interpretable by design, such as differential equations and physics-based simulations in fields such as physics, meteorology, and ecology.

If you value interpretability, you might decide that you only use machine learning algorithms that produce interpretable models. You would still approach the modeling task in a typical performance-first manner, but restrict our solutions to be only models that you deem interpretable. It would still be about optimization, but you strongly limit the hypothesis space, which is the pool of models that are deemed okay to be learned.

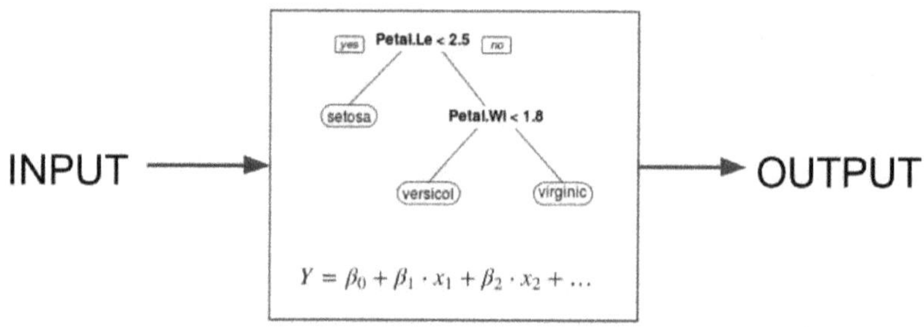

Figure 9.1.: Some models are (supposedly) interpretable by design.

Interpretable machine learning models may include:

- Anything with linearity. Essentially the entire pantheon of frequentist models. Linear regression, logistic regression, frameworks such as generalized linear models (GLMs) and generalized additive models (GAMs), ...
- Decision trees. Structures that present the prediction in the form of typical but not necessarily binary trees that are split by features.
- Decision rule lists
- Combinations of linear models and decision rules such as RuleFit [91] and model-based trees [92]
- Case-based reasoning

Researchers continue to invent new machine-learning approaches that produce models that are interpretable by design [93], but there's also controversy as to what even constitutes an interpretable model [94]. Arguably, not even a linear regression model is interpretable if you have many features. Others argue that whether a particular model can be seen as interpretable depends on the audience and the context. But whatever your definition of interpretation: By restricting the functional form of the models, it becomes functionally better controllable. Even if you can argue that linear regression isn't as easy to interpret, then it is still true that the model contains no interactions (at least on the input space of the model). And models that are interpretable by design, or at least structurally simpler, can help with goals such as discovery, improvement, and justification.

- **Discover:** A logistic regression model allows insights about the features that increase the probability of, e.g., diabetes.
- **Improve:** Regression coefficients with a wrong direction may signal that a feature was wrongly coded.
- **Justify:** A domain expert can manually check decision rules.

Interpretability by design has a cost: Restricting the hypothesis may exclude many good models and you might end up with a model that is worse in performance. The winners in machine learning competitions are usually complex gradient-boosted trees (e.g. LightGBM, catboost, xgboost) for tabular data and neural networks (CNN, transformer) for image and text data. It is not the models that are interpretable by design that are cashing in the price money.

Besides lower performance, interpretability-by-design has a conceptual problem when it comes to discovery (inference). To extend the interpretation of the model to the real world, you have to establish a connection between your model and the world. In classical statistical modeling, for example, you make lots of assumptions about what the data-generating process might be. What's the distribution of the target given the features? What's the process of missing values? What correlation structures do you have to account for? Other approaches such as physics-based simulations also have a link between the model and the world: it is assumed that certain parts of the simulation represent parts of the world. For interpretability by design, there is no such link. For example, there is no justification for saying that almond yield is produced by decision tree-like processes.

Another problem is performance: What do you do when you find models that have better predictive performance than your interpretable one? A model that represents the world well should also be able to predict it well. You would have to argue why your interpretable model better represents the world, even though it predicts it less well. You might have just been oversimplifying the world for your convenience. This weakens any claims to link the model to reality. An even worse conceptual problem is the Rashomon effect [95].

> **i** Origin of the name Rashomon
>
> Rashomon is a 1950 Japanese film. It tells the story of a murdered samurai from four different perspectives: a bandit, the samurai's wife, the samurai's spirit communicating through a psychic, and a commoner observing the event. While each perspective is coherent in itself, it is incompatible with the other three stories. This so-called Rashomon effect has become an established concept in law, philosophy, and, as you can see, statistics and machine learning.

The Rashomon effect describes the phenomenon that multiple models may have roughly the same performance, but are structurally different. A head-scratcher for interpretation: If optimization leads to a set of equally performant models with different internal structures, then you can't pick one over the other based purely based on an argument of performance. An example: Decision trees are inherently interpretable (at least if short) but also inherently unstable – small changes in the data may lead to very different trees even if the performance might not suffer too much. If you use a decision tree for discovery, then the Rashomon effect makes it difficult to argue that this is the exact tree that you should even be interpreting.

The verdict: Inherently interpretable models are much easier to justify than their more complex counterparts. However, using them for insights (inference) has conceptual problems. Fortunately, there's also the class of post-hoc interpretation methods, for which we have a theory of how their interpretation may be extended to the modeled phenomenon.

9.3. Model-specific post-hoc interpretability

Instead of opposing model complexity, you can also embrace it and try to extract insights from the complex models. These interpretability methods are called post-hoc, which translates to "after the event", the event being model training. Post-hoc methods can be either model-specific or model-agnostic: Model-specific interpretation methods work with the structure of the model, whereas model-agnostic methods treat the model as a black box and only work with the input-output data pairs. Model-specific methods are tied to a specific model type and require that you inspect the model, for example:

- Gini importance leverages the splitting criterion in decision trees to assign importance values to features
- Transformers, a popular neural network architecture, has an attention layer that decides which part of the (processed) input to attend to for the prediction. Attention visualization is a model-specific post-hoc approach to interpret transformers.
- Activation maximization approaches assign semantic meaning to individual neurons and layers: What concept maximally activates each neuron?
- Gradient-based explanation methods like Grad-CAM or layer-wise relevance propagation (LRP) make use of neural network gradients to highlight the inputs that affected the predictions strongly

Model-specific methods occupy an odd spot in the interpretability space. The underlying models are usually more complex than the interpretability-by-design ones and the interpretations can only address subsets of the model. For most interpretation goals, model-specific post-hoc interpretation is worse than interpretability-by-design.

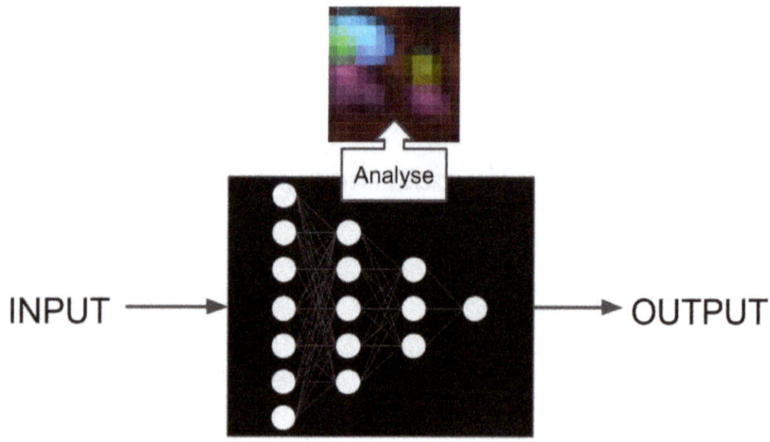

Figure 9.2.: Model-specific post-hoc interpretation analyze parts of the model.

When it comes to inference, model-specific methods share the same conceptual problem with interpretability by design: If you want to interpret the model in place of the real world, you need a theory that connects your model with the world and allows a transfer of interpretation. Model-specific approaches, as the name says, rely on the specific model they are building, and such a link between model and reality would require making assumptions about why this model's structure is representative of the world, which are probably invalid [96]. However, they still may be useful in pointing out associations for forming causal hypotheses (see Chapter 10).

The line is blurred between interpretability by design and model-specific post-hoc interpretation. You might see a linear regression model as inherently interpretable because the coefficients are directly interpretable as feature effects. But what if you log-transform the target? Then you also have to transform the coefficients for interpretation. You can argue that this is still interpretable by design, but you can add many modifications that make the model stray further from interpretability heaven. And for interpretation, you rely more and more on post-hoc computations, like transforming the coefficients, visualizing splines, etc.

Now that we are done talking down on model-specific interpretation, let's talk about our favorite methods: model-agnostic interpretation techniques. And yes, we make no secret out of this, but we are quite the fans of model-agnostic interpretation.

9.4. Model-agnostic post-hoc interpretation methods

You are tasked to write a manual for the world's weirdest vending machine. Nobody knows how it works. You tried to open it but decided against inspection for fear of breaking the machine. But you have an idea of how you could write that manual. You start pressing some of the buttons and levers until, finally, the machine drops an item: a pack of chips with the flavor "Anchovies". Fortunately, they are past the due date. You feel no remorse throwing them away. The other good news is that you made the first step toward writing that manual. The secret: Just try things out systematically and find out what happens.

The fictive almond yield model is just like the vending machine. By intervening in the input features, the output (aka the prediction) changes and you can gather information about the model behavior. The recipe is to sample data, intervene in the data, predict, and aggregate the results [97]. And that's why model-agnostic interpretation methods are post-hoc – they don't require you to access the model internals or change the model training process. Model-agnostic interpretation methods have many advantages:

- You can use the same model-agnostic interpretation method for different models and compare the results.

- You are free to include pre- and post-processing steps in the interpretation. For example, when the model uses principal components as inputs (dimensionality reduction), you can still produce the interpretations based on the original features.
- You can apply model-agnostic methods also to inherently interpretable models like using feature importance on decision trees.

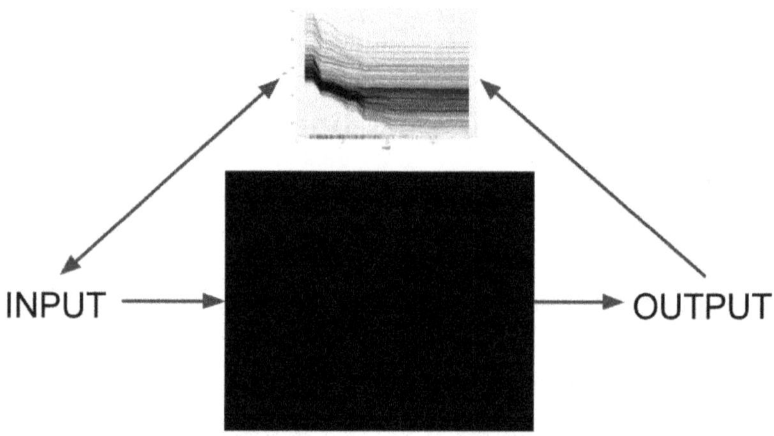

Figure 9.3.: Model-agnostic interpretation ignores the inner workings of the model and studies input-output pairs.

One of the simplest model-agnostic to explain is permutation feature importance: Imagine the almond yield researcher wants to know which features were most important for a prediction. First, they measure the performance of the model. Then they take one of the features, say the amount of fertilizer used, and shuffle it, which destroys the relationship between the fertilizer feature and the actual yield outcome. If the model relies on the fertilizer feature, the predictions will change for this manipulated dataset. For these predictions, the researchers again measure the performance. Usually, shuffling makes the performance worse. The larger the drop in performance, the more important the feature. Figure 9.4 shows an example of permutation feature importance from the almond yield paper [9].

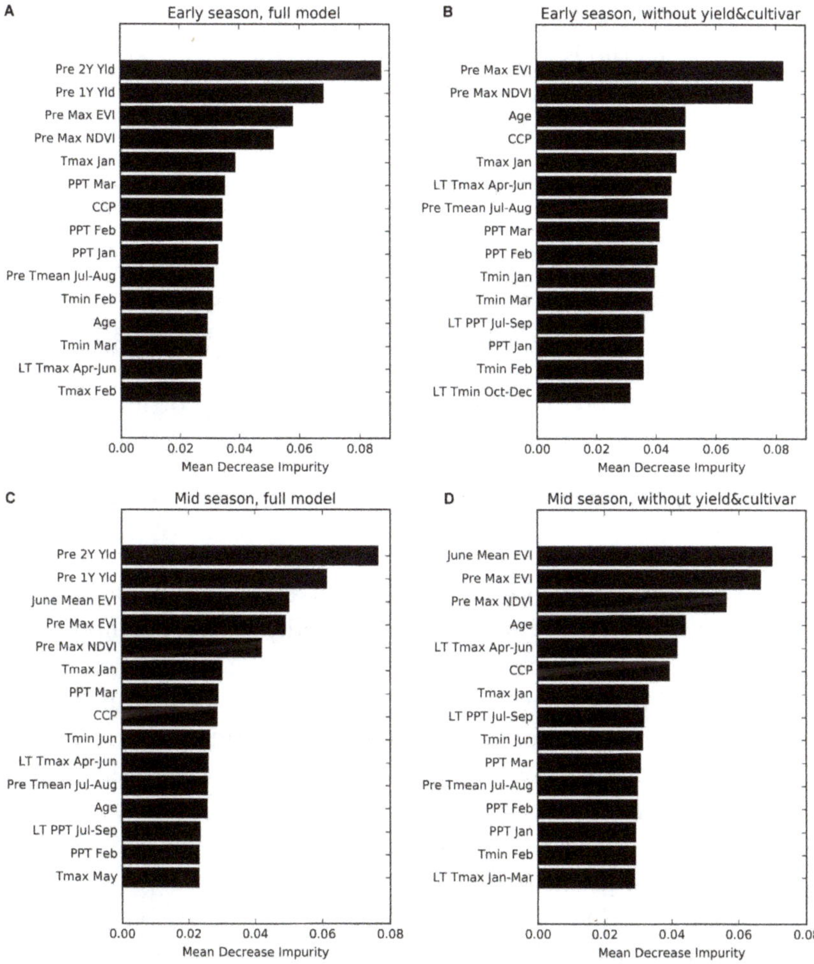

Figure 9.4.: Feature importances for almond yield. Figure by Zhang, Jin, Chen, and Brown (2019) [9], CC-BY (https://creativecommons.org/licenses/by/4.0).

PFI is one of many model-agnostic explanation methods. This chapter won't introduce all of them, because that's already its own book called Interpretable Machine Learning[2] [98] and you can read

[2] https://christophm.github.io/interpretable-ml-book/

it for free! But still, we'll provide a brief overview of the interpretability landscape of model-agnostic methods. The biggest differentiator is local versus global methods:

- Local methods explain individual predictions
- Global methods interpret the average model behavior.

9.4.1. Local: Explaining individual predictions

The almond researchers might want to explain the yield prediction for a particular field and year. Why did the model make this particular prediction? Explaining individual predictions is one of the holy grails of interpretability. Tons of methods are available. Explanations of predictions attribute, in one way or another, the prediction to the individual features.

Here are some examples of local model-agnostic interpretation methods:

- Local surrogate models (LIME) [99] explain predictions by approximating the complex model locally with a neighborhood-based interpretable model.
- Scoped rules (anchors) [100] describe which feature values "anchor" a prediction, meaning that within this range, the prediction can't be changed beyond a chosen threshold.
- Counterfactual explanations [101] explain a prediction by examining which features to change to achieve a desired counterfactual prediction.
- Shapley values [102] and SHAP [103] are attribution methods that assign the prediction to individual features based on game theory.
- Individual conditional expectation curves [104] describe how changing individual features changes the prediction.

9.4.2. Global: Interpreting average model behavior

While local explanations are about data points, global explanations are about datasets: they describe how the model behaves, on average, for a given dataset, and, by extension, the distribution that the dataset represents.

We can further separate global interpretability:

- **Feature importance methods** [105] rank features by how much they influence the predictions. Examples:
  - Permutation feature importance ranks features by how much permuting the feature destroys the information.
  - SAGE [106] ranks features based on predictive performance when the model is retrained.
  - SHAP importance [103] ranks features based on the average absolute SHAP values.
- **Feature effect methods** describe how features influence the prediction. We can further divide feature effect methods into main and interaction effects: Main feature effects describe how an isolated feature changes the prediction, as shown in Figure 9.5. Interaction effects describe how features interact with each other to influence the predictions. Examples:
  - Partial dependence plots [107]
  - Accumulated local effect plots [108]
  - SHAP dependence plots [103]

We see global interpretation methods as central, especially for discovery.

9.4.3. Interpretation for scientific discovery

How useful is model-agnostic interpretation for scientific insights? Let's start with two observations:

- Interpretation is first and foremost about the model.
- If you want to extend the model interpretation to the world, you need a theoretical link.

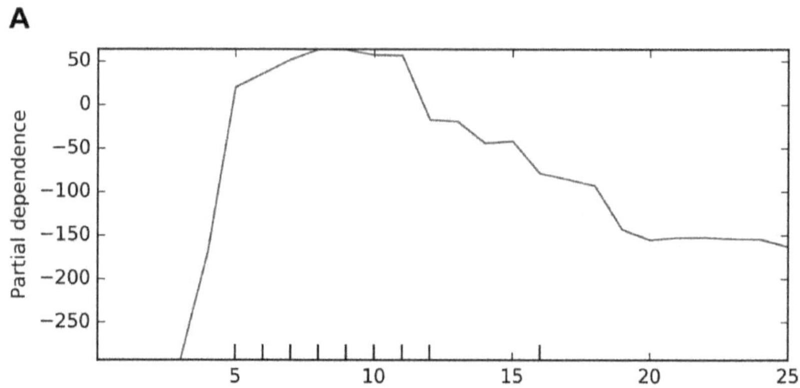

Figure 9.5.: Feature effect using PDPs for "age of orchard" on almond yield. Figure by Zhang, Jin, Chen, and Brown (2019) [9], CC-BY (https://creativecommons.org/licenses/by/4.0).

Let's say you model the almond yield with a linear regression model. The coefficients of the linear equation tell you how the features linearly affect the yield. Without further assumption, this interpretation concerns only the model. In classical statistical modeling, a statistician might make the assumption that the conditional distribution of yield given the features is Gaussian, that the errors are homoscedastic, that the data are representative, and so on, and only then carefully make a statement about the real world. Such results in turn may have real-world consequences, such as deciding the fertilizer usage.

In machine learning, you don't make these assumptions about the data-generating process but let predictive performance guide modeling decisions. But can we establish a theoretical link between the model interpretation and the world maybe differently? We've discussed that establishing this link isn't easy when it comes to model-specific interpretation:

- You would have to justify whether the specific model structures represent the phenomenon you study. That's unreasonable for most model classes: What reason do you have to believe that phenomena in the world are structured like a decision rule list or a transformer model?

- The Rashomon effect presents you with an unresolvable conflict: If different models have similar predictive performance, how can you justify that the structure of one model represents the world, but the others don't?

But there's a way to link model and reality via model-agnostic interpretation. With model-agnostic interpretation, you don't have to link model components to variables of reality. Instead, you interpret the model's **behavior** and link these interpretations to the phenomenon [52]. This, of course, also doesn't come for free.

- You have to assume there is a true function f that describes the relation between the underlying features and the prediction target (see Section 12.5.2 for theoretical background on this true function).
- The model needs to be a good representation of function f or at least you should be able to quantify uncertainties.

Let's make it more concrete: The almond researcher has visualized a partial dependence plot that shows how fertilizer use influences the predicted yield. Now they want to extend the interpretation of this effect curve to the real world. First, they assume that there is some true f that their model $\hat{f}$ tries to approximate, a standard assumption in statistical learning theory [109].

The partial dependence plot is defined as:

$$PDP(x) = \mathbb{E}[\hat{f}(x_j, X_{-j})]$$

and estimated with:

$$\frac{1}{n}\sum_{i=1}^{n} \hat{f}(x_j, x_{-j}^{(i)})$$

The index indicates the feature of interest, and $-j$ indicates all the other features. You can't estimate the true PDP because f

is unknown, but you can define a theoretical partial dependence plot on the true function by replacing the model $\hat{f}$ for the real f: $PDP_{true}(x) = \mathbb{E}[f(x_j, X_{-j})]$.

That allows you, at least in theory and simulation, to compare true PDP with estimated PDP, as visualized in Figure 9.6.

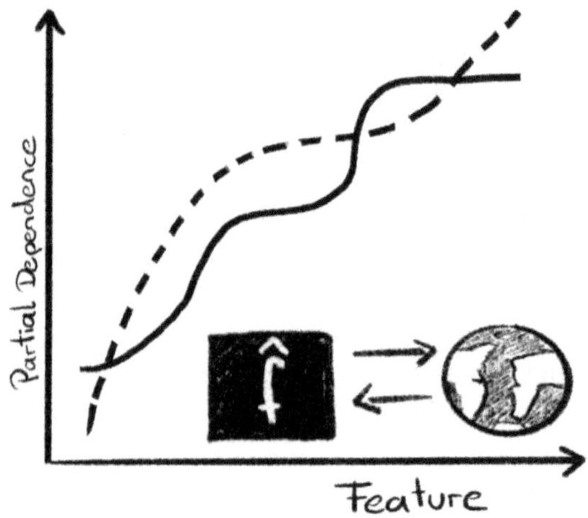

Figure 9.6.: Assumption: there is a "true" PDP of the data-generating process.

That's what we have done in [110] and discussed more philosophically in [52].[3] We'll give an abbreviated sketch of our ideas here. The error between the true and the model PDP consists of 3 parts:

- The *model bias* describes the bias of your learning algorithm.
- The *model variance* describes the algorithm's variance over datasets from the same distribution.
- The *estimation error* describes the variance that arises from estimating the PDP empirically.

[3]Our work on extending interpretation for insights is what motivated us to write the book you are reading.

Ideally, you can either reduce or remove each source of uncertainty or at least quantify them. The model bias is the toughest part: You can't easily quantify it because it would require you to know the true function f. A way to reduce the bias is to train and tune the models well and then assume that the model's bias is negligible, which is a strong assumption to make, of course. Especially considering that many machine learning algorithms rely on regularization which may introduce a bias to reduce variance. An example of model bias: If you use a linear model to model non-linear data, then the PDP will be biased since it can only model a linear relation.

The model variance is a source of uncertainty that stems from the model itself being a random variable since the model is a function of the training data, which is merely a sample of the distribution. If you would sample a different dataset from the same distribution then the trained model might come out slightly different. If you retrain the model multiple times with different data (but from the same distribution), you get an idea of the model's variance. This makes model variance at least quantifiable.

The third uncertainty source is the estimation error: The PDP, like other interpretation methods, is estimated with data, so it is subject to variance. The estimation error is the simplest to quantify since the PDP at a given position x is an average for which you know the variance.

In our paper, we showed that permutation feature importance also has these 3 sources of uncertainty when comparing model PFI with the "true" PFI. We have conceptually generalized this approach to arbitrary interpretation methods in [52]. While we haven't tested the approach for all interpretation methods in practice, our intuition says it should be similar to PDP and PFI. This line of research is rather new and we have to see where it leads. To justify using machine learning + post-hoc interpretation as an inference about the real world this alone might be too thin. But then again, many researchers already use machine learning for insights, so having **some** theoretical justification is great to have.

There's one challenge to interpretability that gives us bad dreams, specifically for the goal of insights: correlated features.

9.5. Correlation may destroy interpretability

When features correlate with each other, model-agnostic interpretation may run into issues. To understand why, let's revisit how most model-agnostic methods work: Sample data, intervene in the data, get model predictions, and aggregate the results. During the intervention steps, new data points are created, often treating features as independent.

For example, to compute the permutation importance of the feature "absolute humidity" for the almond yield model, the feature gets shuffled (permuted) independently of the other features such as temperature. This can result in unrealistic data points of high humidity but low temperatures. These unrealistic new data points are fed into the model and the predictions are used to interpret the model. This may produce misleading interpretations:

- The model is probed with data points from regions of the feature space where the model doesn't work well (because it was never trained with data from this region) or even makes extreme predictions.
- Unrealistic data points shouldn't be used for the interpretation, as they aren't representative of the studied reality.
- Features don't independently change in the real world

Given these issues, why do many still use so-called "marginal" methods for interpretation that treat features as independent? Two reasons: 1) It is technically much easier to, for example, shuffle a feature independently compared to shuffling the feature in a way that preserves the correlation. 2) In an ideal world, you want a disentangled interpretation where you can study each feature in isolation.

In practice, you can avoid or at least reduce the problem of correlated features with these approaches:

- Study correlations between features. If they are low, you can proceed with the usual marginal version of the interpretation methods.
- Interpret feature groups instead of individual features. For most interpretation methods you can also compute the importance or effect for an entire group of features. For example, you can compute the combined PFI for humidity and precipitation by shuffling them together.
- Use conditional versions of the interpretation methods. For example, conditional feature importance [111], conditional SHAP [112], M-Plot and ALE [108] and subgroup-wise PDP and PFI [113], Leave-one-covariate out (LOCO) [114].

The third option, conditional interpretation, is more than a technical fix.

Let's say you want to shuffle the humidity feature to get its importance for almond yield. It is correlated with temperature, so you have to ensure that new data points respect the correlation structure. Instead of shuffling humidity independently, you sample it based on the values of temperature. The first data point has a high absolute humidity, you draw the temperature feature based on this high humidity. This leads to the "permuted" humidity feature to respect the correlation with precipitation, while breaking the relation with the yield target.

Conditional intervention changes the interpretation. For example, for permutation feature importance, the interpretation changes to: Given that the model has access to temperature, how important is it to know the humidity in addition? The more complex the dependence structures between features are, the more complex the (conditional) interpretation becomes. To use conditional interpretation, you must understand how the interpretation changes compared to the marginal versions. But on the other side, we'd argue that conditional interpretation is the way to go, especially for the goal of insights. Not only because it fixes the extrapolation problem, but also because for insights you are interested in how the features are related to the target regardless of the model.

9.6. Interpretability is just one part

Model interpretability alone might not solve your problems. Rather, it is strongly interlinked with the other topics covered in this book.

- Interpretability for scientific insights is tightly linked to generalization (see Chapter 7). If you aren't aware of how the data were collected and which population the dataset represents, it is unclear what to do with the conclusions from model interpretation.
- When you justify the model for your boss, your colleague, or your peers, it is by showing the model coheres with domain knowledge (see Chapter 8).
- Understanding causality (see Chapter 10) is crucial for interpretation. For example, a feature effect may be rather misleading if confounders are missing from the model.

10. Causality

Machine learning exploits all available information to make predictions – and there are many ways to predict things. A good example of this is assessing a person's COVID risk. You can predict it from features such as:

- wearing a mask
- symptoms like dry cough
- vitamin D levels

From a causal perspective, these features relate differently to the COVID risk. This difference is illustrated in the so-called *causal graph* below, which is a graphical tool for thinking about causal dependencies.

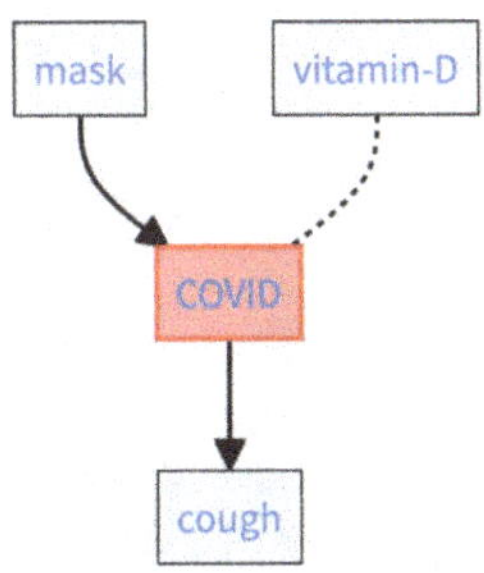

- **Causes:** Wearing a mask is a causal factor for the COVID risk. Why is it a cause? Well, those who put on a mask can thereby lower their COVID risk [115]. The fact that masks are causal for COVID risk is highlighted by the outgoing arrow from the mask node to the COVID node.

- **Effects:** Dry cough is an effect of COVID [116]. People who are healthy and then get infected with COVID likely get dry cough as a symptom. Importantly, it is not a cause of COVID – if people take cough sirup this may help against their cough but does not change their COVID risk. The fact that dry coughing is an effect of COVID is highlighted by the incoming arrow to the cough node from the COVID node.
- **Associations:** Vitamin D levels are associated with COVID risk [117]. For example, in the overall population, people with higher vitamin D levels less often get COVID than people with low vitamin D levels. Still, it is unclear if higher vitamin D levels cause a lower COVID risk or if the association arises for entirely different reasons. The unexplained association between vitamin D levels and COVID is highlighted by an undirected dashed arrow.[1]

Initially, there was a lot of hype about machine learning in Raven Medicine. Some even proclaimed the end of costly medical experiments. But the enthusiasm soon waned: Machine learning provided little insight into how to treat Ravens to make them healthy. Was the red pill or the blue pill more effective? The old proverb "correlation is not causation" hit the machine learning enthusiasts hard.

[1]There is also an association between COVID risk and the FIFA football ranking [118]. Here, the causal story is spuriously Messi...

10.1. Prediction does not require causal understanding

All the features we have just listed can be equally helpful in predicting COVID risk with machine learning. You may decide to rely only on causes or only on effects:

- **Predict effects from causes:** An example is the protein folding problem discussed at the beginning, where the amino acid sequence causally determines the protein structure [119].
- **Predict causes from effects:** An example is predicting the presence of a black hole from its gravitational effects on the surrounding bodies.
- **Mixed prediction:** This is probably the most common case. An example is medical diagnosis. To diagnose malaria, doctors take into account symptoms like high fever but at the same time causes such as a mosquito bite from your South America travels.

In the end, it is your decision as a modeler what you want to incorporate in your prediction model, whether it is causes, effects, causes of effects, or even spurious associations. For the machine learning model to successfully predict, all that matters is that the feature contains information about the target variable.

10.2. Costly experiments distinguish causes from effects

Controlled experiments are the best approach to distinguish between causes and effects. Let's say you have two variables, vitamin D and COVID. To establish whether higher vitamin D levels reduce the COVID risk, you could run a *randomized control trial (RCT)*. You have 10,000 test subjects: 5,000 *randomly* selected subjects belong to the treatment group (they get vitamin D supplements); the other 5,000 belong to the control group (they get placebos). If after a certain time, the COVID infections in the treatment group are significantly lower than in the control group, you can conclude that vitamin D is a causal factor for COVID risk. Easy, right?!

Unfortunately, conducting controlled experiments in the real world is cumbersome. It takes a lot of time, money, and resources. Some experiments can be ethically or legally problematic, such as the administration of harmful drugs to humans. Other experiments are beyond the capabilities of humans – like investigating the effects of pushing Jupiter out of its solar trajectory – you'd need to be as strong as Saitama [2] for that...

Social scientists, in particular, often deal with *observational data*. Observational data is data that is measured without controlled conditions. Like asking random people on the streets for their vitamin D levels and if they had COVID. Observational data is what we gather all the time – the little measuring device in your pocket is piling up tons of it.

10.3. Machine learning can generate causal insights

Observational data is what you are likely feeding your machine learning algorithm. Unfortunately, observational data alone does not provide causal insight [17].

> **i Insight**
>
> It is impossible to distinguish causes from effects from observational data alone [17].

If you cannot even distinguish between causes and effects from observational data alone, how can machine learning help with causality? It turns out that machine learning can help with causal inference (answering causal questions with the help of data), but

[2] Saitama is the protagonist of the anime series "One-Punch Man" who defeats his opponents with just one punch. In one of the episodes, he accidentally damages good old Jupiter.

usually not for free: You have to make assumptions about causal structures.

These are the causal questions that machine learning can help with:

- **Studying associations to form causal hypotheses:** Machine learning helps to investigate associations in data such as the association between COVID and vitamin D levels, which can be the starting point for causal hypothesizing.
- **Estimating causal effects:** Machine learning can be used to estimate causal effects, such as quantifying the effect of vitamin D on the COVID risk.
- **Learning causal models:** You can use machine learning to learn causal models. Causal models are formal tools to reason about real-world interventions and counterfactual scenarios.
- **Learning causal graphs:** Causal graphs encode the direction of causal relationships, and machine learning can help learn them directly from data.
- **Learning causal representations:** Machine learning can learn causal variables, which are high-level representations (e.g. objects) of low-level inputs (e.g. pixels).

These five tasks structure the rest of this chapter.

10.4. Studying associations to form causal hypotheses

You may have heard that machine learning models are capturing complex associations in data. But what are associations? It is easiest to understand associations by their opposite, namely statistical independence. Two events are independent if knowing one event is uninformative about the other, e.g. knowing about your COVID risk does not tell anything about the weather on the planet Venus. More formally, two features A and B are *statistically independent* if $\mathbb{P}(A \mid B) = \mathbb{P}(A)$. We call two features *associated* whenever they are not statistically independent. For example, COVID risk is associated with wealth. Statistically, the higher your wealth the lower your COVID risk [120]. But be careful, associations are complex creatures. For example, wealthier venders are likely to have a higher COVID risk than poor venders because they have contact with more customers. This is called an *interaction effect*.

Like classical statistical techniques, machine learning together with interpretability methods (see Chapter 9) enables you to read out complex properties of your data. You can for example study:

- **Feature effects & interactions:** How is the target associated with certain input features?
- **Feature importance:** How much do certain features contribute to the prediction of the target?
- **Attention:** What features is our model listening to when predicting the target?

10.4.1. Form causal hypotheses with the Reichenbach principle

Assume you find in your data the association between vitamin D and the COVID risk. Then, the **Reichenbach principle** can guide you in forming a causal hypothesis. The principle states three possibilities where the association can come from [121]:

1. Vitamin D can be a cause of COVID.

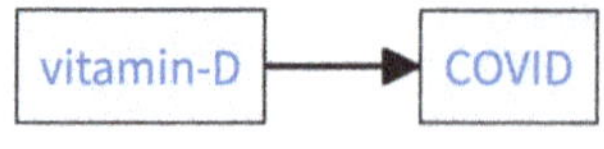

2. COVID can be a cause of vitamin D.

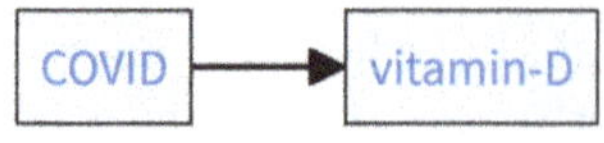

3. Or, there is a common cause Z of both, vitamin D and COVID.

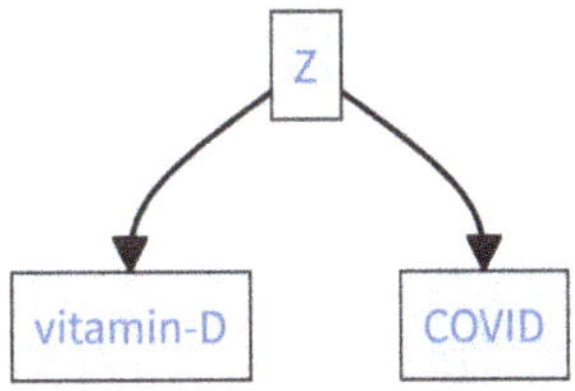

Nothing against Reichenbach, but in reality, there are two more explanations for the association:

- First, the association can be *spurious*. It is only there because of limited data. If you continue gathering data, the association will vanish. Figure 10.1 shows an almost perfect association between worldwide non-commercial space launches and the number of doctorates awarded in sociology [122]. But is it causal? Perhaps Elon Musk can boost his business by awarding a few PhD scholarships in sociology...
- Second, the association can result from a *selection bias* in the data collection process. It occurs if the data sample is not representative of the overall population. For example, while niceness and handsomeness are probably statistically independent traits in the overall population, they can become dependent if you only look at people you would date [123].

But how to identify the correct explanation for the association between COVID and vitamin D? Because there is a lot of data (not spurious) and the association is found in the overall population (no selection bias), you may infer that one of Reichenbach's three cases applies. Gibbons et al. [124] compared a large group of veterans receiving different vitamin D supplementations to a group without treatment. They found that the treatment with vitamin D reduces the risk of getting COVID significantly and therefore recommend broad supplementation. Even though the study was not a randomized control trial, it strongly indicates that there is a causal link between vitamin D levels and COVID risk.

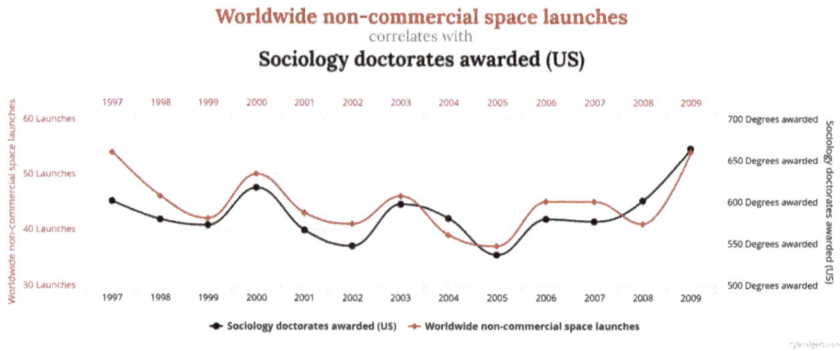

Figure 10.1.: There is an almost perfect association between the worldwide non-commercial space launches and the number of sociology doctorates awarded. Chart by Tyler Vigen, CC-BY (https://creativecommons.org/licenses/by/4.0/)

The scientific story may continue at that point: There might be unknown common causes of both vitamin D and COVID risk. Moreover, the causal link does not explain by what mechanism in the body vitamin D acts on COVID risk. That's how science works, new questions keep popping up...

10.4.2. Select explanatory variables with feature importance

Importances of features primarily allow you to evaluate which features are most informative for predicting the target. For example, they may tell you whether vitamin D levels contain any information about your COVID risk beyond the information that is already contained in the mask and the cough features. But importance itself gives no causal insight. It can still be useful for a causal analysis: importance allows you to understand what information is redundant, noisy, or just unnecessary. This may prove helpful in selecting sparse and robust features for a causal analysis [125].

Imagine you know a person's age, country of origin, gender, favorite color, and favorite band. You want to predict again her COVID risk. By studying feature importance, you may find out that:

- The favorite color has very little importance under all constellations. It might even have negative importance as it introduces noise.
- The favorite band is extremely informative. You can even drop all other features without a major loss in predictive performance.
- If you know the age, country of origin, and gender of a person, knowing her favorite band does not add much.

Depending on your goal, this can be very insightful. You can generally drop the favorite color feature, it is not predictive and will not provide insights for a causal analysis. If you want to reason about policy interventions to lower the COVID risk, you can exclude the favorite band from your causal analysis. If, on the other hand, you are a sociologist who wants to study how taste in music is related to health, you may want to keep all features.

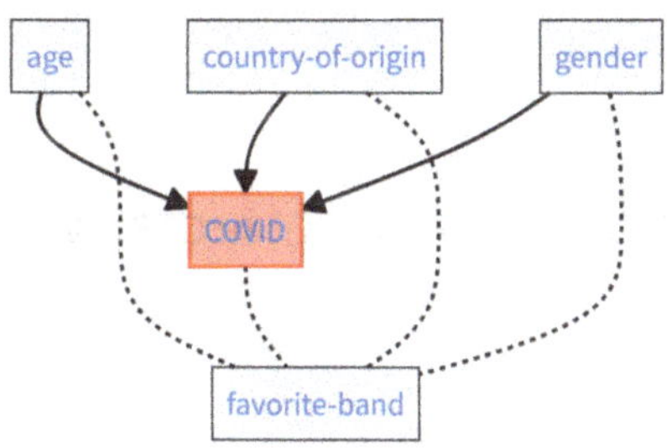

10.4.3. Find high-level variables with attention

The question of what a model is attending to in its decisions is prominent in interpretability research, particularly in image classification [126]. Although scientists may be more concerned with the relationship in the data than in the model, model attention may still be relevant if the input features have meaning only in the aggregate, not individually. Think of pixels that constitute images, letters that constitute sentences, or frequencies that constitute sounds. Attention can point you to meaningful constructs or subparts that are associated with the target and allow you to form causal hypotheses about them.

For example, while predicting COVID from CT scans sounds like taking a sledgehammer to crack a nut, it can be an interesting approach to learning more about how the disease affects the human body. Even though the input features themselves are pixels, they collectively form higher-level concepts that can be highlighted by saliency-based interpretability techniques. This approach was for example used by [127] to detect lesions in the lungs of COVID patients as illustrated in Figure 10.2.

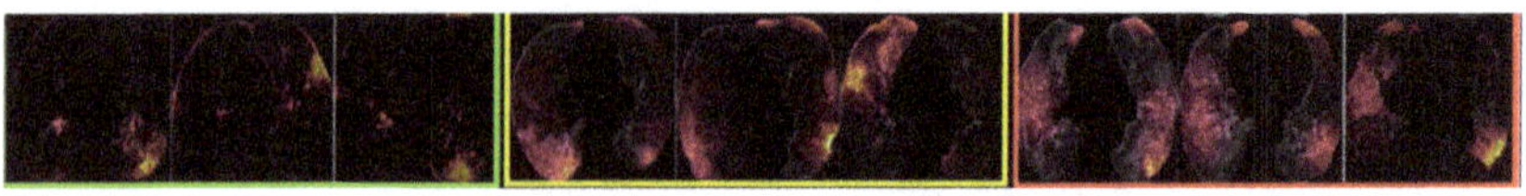

Figure 10.2.: Lesions in CT scans of lungs detected with integrated gradients, small in green, mixed in yellow, large in red. Figure by [127], CC-BY (https://creativecommons.org/licenses/by/4.0/)

10.5. Estimating causal effects with machine learning

So far, we have assumed that researchers are searching in the dark, where they use machine learning to find interesting associations to form and test causal hypotheses. However, often researchers have causal knowledge or intuitions about how features relate to each other. Their goal is causal inference, or, to be more precise, to estimate causal effects, which is a more well-defined problem than just exploring associations.

Causal graphs show causal dependencies, e.g. if you put on a mask, it will causally affect your COVID risk. But this begs the question of **how** it affects your risk. Is wearing a mask lowering your COVID risk and if yes, by how much? This question asks for a *causal* *effect.* Questions for causal effects were omnipresent during the pandemic:

- How does vaccination affect the COVID risk of elderly people?
- How does the average COVID risk change with contact restrictions compared to without them?
- How does treatment with vitamin D pills lower the COVID risk of African Americans [124]?

Causal effects always have the same form. You compare a variable of interest such as COVID risk *with treatment*, for example, masks, against those *without treatment* (no-mask). This gives us the so-called *Average Treatment Effect* (ATE). In the case of masks being the treatment, we can formally describe the effect on the COVID risk:

$$\text{ATE} = \mathbb{E}[\text{COVID} \mid do(\text{mask} = 1)] - \mathbb{E}[\text{COVID} \mid do(\text{mask} = 0)]$$

Let's unpack this formula. The *do* in the ATE denotes the do-operator, which means you intervene on a variable and set it to a certain value. Thus, the first term describes the expected COVID risk if you force people to wear masks. The second term describes the expected COVID risk if you force people not to wear masks. The difference between the two terms describes the causal effect of masks on the COVID risk.

In many cases, rather than asking for the ATE, you ask for the treatment effect for a specific subgroup of people that share certain characteristics. During the pandemic, for example, researchers might have asked about the causal effect of masks for children under 10 years of age. This is the so-called *Conditional Average Treatment Effect* (CATE):

$$\begin{aligned}\text{CATE} = \;&\mathbb{E}[\text{COVID} \mid do(\text{mask} = 1), \text{age} < 10] \\ &- \mathbb{E}[\text{COVID} \mid do(\text{mask} = 0), \text{age} < 10]\end{aligned}$$

The *do()* terms in the two formulas give the impression that you have to run controlled experiments to estimate the causal effects. And conducting experiments would be the most reliable way – if you can do it, you should! Unfortunately, experiments are sometimes not feasible. In some situations, you are lucky because you can estimate the ATE and CATE from observational data alone. If this is the case, the causal effect is *identifiable.*

10.5.1. Identify causal effects with the backdoor criterion

The classical way to identify a causal effect is via the so-called *backdoor criterion.* The idea is simple. Let's say you want to know the ATE of masks on the COVID risk. You only have observational data. But your data allows you to read out: 1. how many people with masks got infected and 2. how many people without masks got infected. Just subtract the latter from the former and you are finished, right? No! The problem is that the information you want to obtain is potentially blurred by other factors. For example:

- **Sticking to the rules:** People who stick to the rules are more likely to wear masks and also follow other rules like contact restrictions. Therefore, they also have on average a lower COVID risk.
- **Number of contacts:** People who have many contacts (e.g. postmen, salespeople, or doctors) wear masks to lower their own risk and the COVID risk of others. But many contacts increase also their COVID risk.
- **Age:** Younger people are less likely to wear masks because they think masks do not look cool or because COVID is less dangerous for young people. At the same time, young people are more resistant to getting infected with COVID [128].

Those factors are called *confounders.* They are both a cause of the treatment and the variable of interest. Imagine you knew all those confounders and accounted for them. Then, you could isolate the association that is solely due to the masks themselves. This is exactly what the backdoor criterion is doing. If you know all confounders W, the ATE can be estimated by:

$$\begin{aligned}\text{ATE} = &\mathbb{E}_W[\mathbb{E}_{\text{COVID}|W}[\text{COVID} \mid \text{mask} = 1, W]] \\ &- \mathbb{E}_W[\mathbb{E}_{\text{COVID}|W}[\text{COVID} \mid \text{mask} = 0, W]]\end{aligned}$$

A great success, there are no do-terms left. But be aware that knowledge of all confounders is a super strong requirement that can rarely ever be met in reality. And not knowing a confounder may bias your estimate significantly. The same happens if you mistake a variable for a confounder that is caused by both the treatment and the variable of interest (this is called a *collider bias*). So conditioning on more features is not always better.

The back-door criterion is not the only option to identify causal effects. There is for example also its friendly sibling, the *frontdoor criterion*. Both, the front and the backdoor criterion are special cases of the do-calculus. Whenever it is possible to identify a causal effect, you can do so using the do-calculus. This book will stay at this surface level on the question of identifiability, if you like math and want to dig deeper, check out [129]. The cool thing is, you don't have to know all of that necessarily to do causal analysis. The identification step can be fully automated given a causal graph, for example, using the Python package *DoWhy* [130].

Let's take a look into two machine learning-based methods that allow for estimating causal effects: the T-Learner and Double Machine Learning [3]. Both are designed for the backdoor criterion setting. That means, in both cases, you assume that you have observed all confounders W, and your causal graph looks like this

[3] There are two general approaches for estimating treatment effects with machine learning, model-agnostic and model-specific techniques. Model-specific techniques are based on a certain class of machine learning models, such as GANITE for neural nets [131] or causal forests for random forests [132]. Model-specific techniques often come with the advantage of valid confidence intervals. Other techniques are model agnostic, which means, you can simply plug in any machine learning model into the estimation. The T-Learner and Double Machine Learning we present here are examples of such model-agnostic approaches. Today, there are tons of approaches to estimating causal effects with machine learning, check out [133] or [134] to learn more.

10.5.2. How to estimate causal effects with the T-Learner

The T-Learner is maybe the simplest approach to estimate the ATE (or with slight modifications CATE) from observational data. The name T-Learner stems from using two different learners to estimate the ATE. The T-learner uses the formula from the backdoor criterion:

$$\begin{aligned}\text{ATE} = \mathbb{E}_W[&\underbrace{\mathbb{E}_{\text{COVID}|W}[\text{COVID} \mid \text{mask} = 1, W]}_{\phi_1}] \\ &- \mathbb{E}_W[\underbrace{\mathbb{E}_{\text{COVID}|W}[\text{COVID} \mid \text{mask} = 0, W]}_{\phi_0}]\end{aligned}$$

The T-Learner fits two machine learning models, $\hat{\phi}_0 : \mathcal{W} \to \mathcal{Y}$ and $\hat{\phi}_1 : \mathcal{W} \to \mathcal{Y}$, where $\hat{\phi}_0$ is estimated only with data where mask $= 0$ and $\hat{\phi}_1$ with data where mask $= 1$. The ATE can then be computed by averaging over different values of W via

$$\text{ATE} \approx \frac{1}{n} \sum_{i=1}^{n} \left(\hat{\phi}_1(w_i) - \hat{\phi}_0(w_i) \right).$$

10.5.3. How to estimate causal effects with Double Machine Learning

Double Machine Learning is an improvement over the T-Learner. Not only is the name much cooler, but the estimation of the ATE is much more sophisticated [135]. The estimate is both unbiased and data-efficient. The estimation process works as follows:

1. **Predict outcome & treatment from controls:** Fit a machine learning model to predict COVID from the confounders W. This gives you the prediction $\widehat{\text{COVID}}$. Next, fit a machine learning model to predict whether people wear masks or not from the confounders W. This gives you the prediction $\widehat{\text{mask}}$. You need to split your data into training and estimation data, the machine learning models should be learned only from the training data.
2. **Estimate outcome residuals from treatment residuals:** Fit a linear regression model to predict $\text{COVID}-\widehat{\text{COVID}}$ from $\text{mask}-\widehat{\text{mask}}$. Then, your estimand of interest is the coefficient β_1. Why? Well, surprisingly, this coefficient describes the treatment effect. But you need quite some math to show this, search for the Frisch-Waugh-Lovell theorem if you want to learn more. The linear regression must be fitted on the estimation data not the training data for models from step 1.
3. **Cross-fitting:** Run steps 1 and 2 again, but this time switch the training data you use to fit the machine learning models in step 1 with the estimation data you use in step 2 for the linear regression. Average your two estimates $\widehat{\beta_1}^1$ and $\widehat{\beta_1}^2$ to obtain your final estimate $\widehat{\beta_1} = (\widehat{\beta_1}^1 + \widehat{\beta_1}^2)/2$.

Why does this strange procedure make sense? If you want to get the details, check out [135]. The key ideas are:

- You want to get rid of the bias in your estimation that stems from regularization in your machine learning model. This is done in steps one and two, exploiting the so-called Neyman orthogonality condition.

- You want to get rid of the bias in your estimation that stems from overfitting to the data you have and at the same time be data efficient. If you would train our machine learning model on the same data on which you estimate the linear coefficient, you would introduce an overfitting bias. Splitting the data into training and estimation data circumvents this pitfall. However, splitting the data just once is less data efficient. Thus, you switch the roles of training and estimation data and compute the average in step three.

If you want to use Double Machine Learning in your research, check out the package `DoubleML` in R [136] and Python [137]. Generally, there are many more methods now to estimate causal effects with machine learning, check out [138], [133] or [134] to get an overview.

10.6. Learning causal models if we know the causal graph

Causal graphs are simple objects that contain:

- **Nodes:** Describe the features that are interesting to you, like the COVID risk, if you wear masks, if you have a dry cough, or your vitamin D levels.
- **Arrows:** describe how the features are causally linked. Like the causal arrow between COVID risk and dry cough.

Usually, you pose two additional assumptions on causal graphs – they should be directed and acyclic. Directed means that each arrow has a start node and end node, not unlike the dashed arrow in the case of vitamin D in the beginning which treats both nodes the same. Acyclic means that it should not be possible to start from a node and return to it while only walking along the directed arrows. If the graph is directed and acyclic, we talk about a DAG – a directed-acyclic graph.

So, causal graphs are a tool to reason about causal relations. But wouldn't it be nice to have *causal models*? Models that once they are constructed allow you to think about a whole range of questions without always running this whole treatment effect estimation process?

What is the effect of masks on the COVID risk? How about both masks and contact restrictions? What if you only look at older people? Imagine all these questions can be answered with one model.

10.6.1. Bayesian causal networks allow to reason about interventions

In causal graphs, we talked about the nodes of the graph, like the mask node, the COVID node, or the cough node. But how to interpret these nodes specifically? The perspective of Causal Bayesian Networks (CBNs) is to see these nodes as random variables. If causal graphs are a combination of graphs and causality, CBNs add probability theory to the mix.

CBNs allow you to ask classical probabilistic questions, like what is the probability that people wear masks if they have a cough $\mathbb{P}(\text{mask} \mid \text{cough} = 1)$. But what CBNs are ultimately designed for is answering causal questions about interventions. Like what is the average COVID risk if everyone must wear a mask $\mathbb{P}(\text{COVID} \mid do(\text{mask} = 1))$?

Let's check out this simplified causal graph with only three variables: mask, COVID risk, and cough. Mask is causal for COVID risk and COVID is causal for cough. In this simple setting, you must specify three things to obtain a CBN:

1. The marginal probability of wearing a mask, i.e. $\mathbb{P}(\text{mask})$.

2. The COVID risk for both people who either wear or don't wear masks, that means $\mathbb{P}(\text{COVID} \mid \text{mask} = 0)$ and $\mathbb{P}(\text{COVID} \mid \text{mask} = 1)$.
3. The probability of having a cough if the COVID risk is low or high, that means $\mathbb{P}(\text{cough} \mid \text{COVID} = 1)$ and $\mathbb{P}(\text{cough} \mid \text{COVID} = 0)$.

Let's look at a toy example. Assume that 50% of people wear masks. Assume the COVID risk is 80% low / 20% high if people wear a mask and reversed if people do not wear a mask. And, the probability that people have a cough if they have a high COVID risk is 90%, and conversely, 10% if they have a low COVID risk.

Then, you can ask the following question and address it formally:

- **Observational question:** What is the probability of wearing a mask, having a high COVID risk, and not coughing? This can be computed by:

$$\begin{aligned}
&\mathbb{P}(\text{mask} = 1, \text{COVID} = 1, \text{cough} = 0) \\
&= \mathbb{P}(\text{mask} = 1) \times \mathbb{P}(\text{COVID} = 1 \mid \text{mask} = 1) \\
&\quad \times \mathbb{P}(\text{cough} = 0 \mid \text{COVID} = 1) \\
&= 0.5 \times 0.2 \times 0.1 = 0.01
\end{aligned}$$

So this is a pretty unlikely combination. You can also ask a causal question, like:

- **Interventional question:** What is the probability of having a high COVID risk and cough if the policymaker enforces masks? This can be computed by:

$$\begin{aligned}
&\mathbb{P}(\text{COVID} = 1, \text{cough} = 1 \mid do(\text{mask} = 1)) \\
&= \mathbb{P}(\text{COVID} = 1 \mid \text{mask} = 1) \times \mathbb{P}(\text{cough} = 1 \mid \text{COVID} = 1) \\
&= 0.2 \times 0.9 = 0.18
\end{aligned}$$

Let's generalize the ideas of this example. What do you need for a BCN?

- The marginal distribution $\mathbb{P}(X_r)$ of all nodes in the causal graph without incoming arrows. These nodes are called root nodes.
- For all non-root nodes, you need their conditional distribution $\mathbb{P}(X_j \mid pa_j)$ given their direct causes. The direct causes of a node are called parents.

If you have those two ingredients specified, you have a BCN. It allows you to compute both observational and interventional probabilities:

Observational probabilities:

$$\mathbb{P}(X) = \prod_{i=1}^{p} \mathbb{P}(X_i|pa_i)$$

Interventional probabilities:

$$\mathbb{P}(X_{-j} \mid do(X_j = z)) = \prod_{\substack{i \neq j, \\ j \notin pa_i}} \mathbb{P}(X_i \mid pa_i) \prod_{j \in pa_i} \mathbb{P}(X_i \mid pa_i, X_j = z)$$

To calculate interventional probabilities, you set the intervened variable to the desired value wherever it appears in the formula and the corresponding probability that the variable has this value to 1.

All of this is nice theory and toy modeling. But how about the real world where you do not magically have the BCN? Indeed, in practice, you can learn the BCN if you know the causal graph. All that differs between a BCN and a causal graph are marginal and conditional probabilities. If you have data, this is exactly where machine learning can help – estimating conditional probabilities. We generally advise estimating conditional probabilities with non-parametric machine learning models like neural networks or random forests, however, for small data sizes classical parametric approaches like maximum likelihood estimation or expectation maximization might be preferable [139].

10.6.2. Structural causal models allow you to reason about counterfactuals

Interventional questions are often important if we want to act. But what if you want to explain *why* something has happened? Let's say *you* got COVID and want to know why. Was it not wearing a mask? Or, was it that party last week? Or did you catch it in the metro from this guy who was sneezing heavily?

Answering why questions is always the most difficult, but usually also the most interesting. What is needed to answer such questions? You have to think through counterfactual scenarios. You know that you went to the party and you also know that you got COVID. To answer the why question, you need to find out how you got infected. Would you have got COVID if you hadn't gone to the party? Would you have caught COVID if you hadn't spoken to your charming but coughing neighbor at the party?

BCNs are not expressive enough to reason about counterfactual scenarios. Why not? Firstly, BCNs describe causal relations probabilistically on the level of groups of individuals who share certain features. Like the probability of getting COVID if everyone is forced to the party. The causal relationships are learned from data of people who did and of people who did not attend the party. The why question is more specific. It asks why you individually got COVID. Did you get COVID *because* you attended the party?[4] Secondly, the fact that you got COVID is informative about you, it might reveal information about your properties on which there is no data. Like info about your immune system and how resilient it is against COVID. This information should not be ignored.

[4]In the literature, it is differentiated between counterfactuals and actual causes [140].

To answer why questions you need to simulate alternative scenarios, i.e. to perform counterfactual reasoning. A Structural Causal Model (SCM) is a model designed to simulate such alternative scenarios. Instead of describing probabilistic relationships in the data, a SCM explicitly models the mechanism that generates the data.

Let's look at a super simple SCM with just two factors: whether you go to the party or not and whether you have COVID or not. These factors that you have explicit information about are called the endogenous variables.

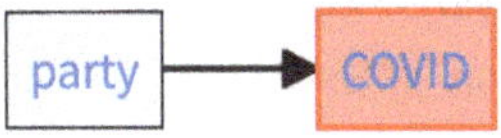

The model moreover contains two *exogenous* variables. Exogenous variables describe background factors on which you have no data but which play a role in determining the endogenous variables. For example, whether you go to the party or not might be determined by whether you are in the mood for partying. Similarly, whether you get COVID or not might be influenced by how well your immune system is currently working. The exogenous variables are very powerful in SCMs. These factors completely determine the endogenous variables.

The SCM is thus specified by:

- Two noise terms: $U_{\text{mood}} = Ber(0.5)$ describes whether you are in the mood for partying and the chances are fifty-fifty. $U_{\text{immune}} = Ber(0.8)$ describes whether your immune system is up and the chances are pretty high (80%) that your immune system works well.
- One structural equation for the endogenous party variable, i.e. party $:= U_{\text{mood}}$. Whether you go to the party is fully determined by whether you are in the mood for it.
- One structural equation for the endogenous COVID variable, i.e. COVID $:= max(0, \text{party} - U_{\text{immune}})$. That means, you

only get COVID if you are at the party and your immune system is down.

This SCM allows you to answer the counterfactual question: You went to the party and you caught COVID, would you have gotten COVID if you hadn't gone to the party? Such counterfactuals are computed in three steps:

- **Abduction:** What does the fact that you got COVID tell you about the noise variables? Well, you could not have caught COVID with your immune system up. Thus, you can infer that $U_{\text{immune}} = 0$. Similarly, you know that you were in the mood for partying because otherwise, you would not have gone, i.e. $U_{\text{mood}} = 1$.
- **Action:** Let's say you would not have gone to the party, that means you intervene and turn the party variable from one to zero, i.e. $do(\text{party} = 0)$.
- **Prediction:** What happens with the COVID variable if you switch party to zero? According to the structural equation of COVID, you only get COVID if both happens, you go to the party and your immune system is down. Thus, you do not get COVID.

In consequence, if you hadn't gone to the party, you would not have caught COVID. So, you got COVID **because** you went to the party.

What if the counterfactual had not been true? Would partying still be a cause? Well, then things get more complicated. It could for instance be that you would have gone for sports instead of the party and caught COVID there. But partying would still have been the actual cause of you catching COVID. Search for *actual causation* to learn more about the topic.

Let's generalize the ideas of the example. For a SCM, you need:

- A set of exogenous variables U that are determined by factors outside the model and independent from each other.
- A set of endogenous variables X that are fully determined by the factors inside the model.

- A set of structural equations F that describe for each endogenous variable X_j how it is determined by its parents and its exogenous variable $X_j = f(pa_j, U_j)$.

Obtaining SCMs in real life is difficult. You may use machine learning to learn the structural equations if you know the causal dependencies between variables. However, even for that, you need quite some data, and making parametric assumptions can be advisable.

Like counterfactuals themselves, SCMs are highly speculative objects. There is no easy way to verify that a given SCM is correct, it always relies on counterfactual assumptions. Still, since humans reason in terms of counterfactuals all the time and are deeply concerned with *why* questions, it is nice to have a formalism that can capture counterfactual reasoning.

10.7. Learning causal graphs from data

Machine learning has revived an old idea – with enough data, we might be able to completely automatize science. In Part 1 and Part 2 of this chapter, you always needed a human scientist in the background. Someone to come up with interesting hypotheses, run experiments, or provide us with a causal graph. This part will be the most ambitious, it asks:

> Can you get a causal graph just from data, without entering domain knowledge? This problem is often referred to as *causal structure learning* or *causal discovery*.

Didn't we tell you in the beginning that observational data alone doesn't do the trick? Correct! But it is intriguing to see how far you can go, especially if you add some additional assumptions. Clear the stage for a truly fascinating branch of research!

Imagine all you have is a dataset containing 5,000 entries. Each entry contains information about a person's BMI, COVID vaccination status, flu vaccination status, COVID risk, fatigue, fever, appetite, and the density of population in the area the person lives in (Example based on [141] and [142]). The task is to structure these features in a causal graph – the problem of *causal discovery*. Let's say the true graph that you want to learn looks like this:

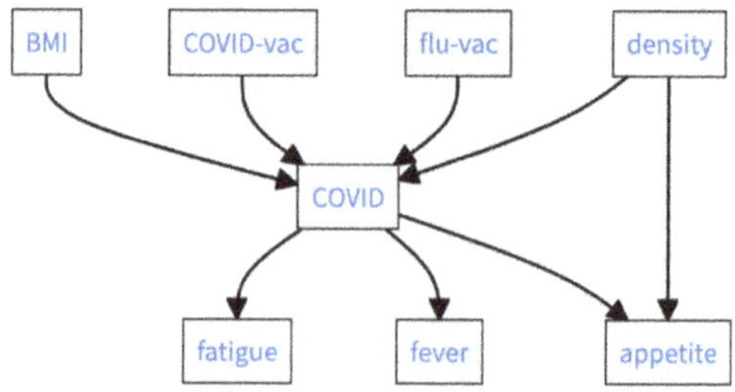

Before you can approach this problem, you have to get some more background on how causal mechanisms, associations, and data relate to each other. The causal mechanism generates data. Within this data, features are associated with each other. That means that causal dependencies induce statistical (in-)dependencies. Particularly, if you have a causal graph, the so-called causal Markov condition applies:

> **i** Causal Markov condition:
>
> Given its parents in the causal graph, a variable is statistically independent of all its non-descendants. Or formally, let G be a causal graph and X_i be any variable in the network, then for all its non-descending variables nd_i holds $X_i \perp\!\!\!\perp nd_i \mid pa_i$. Descendants of a given variable X_i are all the variables that can be reached from X_i by walking along the direction of arrows.

To get an intuition on the causal Markov condition, take a look at the graph below. Consider the variable appetite as an example. Its parents are COVID and density. Its nondescendants are BMI, COVID-vac, flu-vac, fever, and fatigue.[5] The causal Markov condition allows you for example to say that appetite is statistically

[5]The parents are indeed also non-descendants. However, the conditional

independent of fever if you know COVID and density. The formal way to write this is $\text{appetite} \perp\!\!\!\perp \text{fever} \mid \text{COVID}, \text{density}$, where $\perp\!\!\!\perp$ is the symbol for independence.

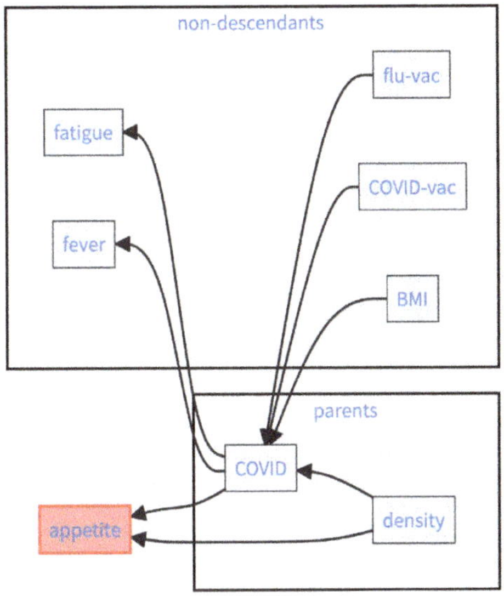

Cool, causal graphs induce statistical independencies, but what to do with that? Well, you have a dataset, so you can test for statistical independencies.

There are many different approaches to test such (conditional) independencies, and they vary regarding the parametric assumptions they pose. Some test the statistical dependency to be Gaussian or linear (e.g. partial correlation), or are tailored for categorical data like the chi-squared test [143]. Others allow for a greater variety of dependencies, like tests based on the Hilbert-Schmidt independence criterion (HSIC) [144]. The current trend goes even more non-parametric towards machine learning based tests for conditional independence. Questions of independence are translated into classification problems, where powerful machine learning models can be utilized (e.g. neural nets or random forests) [145], [111].

independence you can derive from the causal Markov condition between E of C given you know C holds trivially for an arbitrary variable by $\mathbb{P}(E, C \mid C) = \mathbb{P}(E, C, C)/\mathbb{P}(C) = \mathbb{P}(E \mid C) = \mathbb{P}(E \mid C)\mathbb{P}(C \mid C)$.

However, the fewer parametric assumptions you pose, the more statistical dependencies are possible. Thus, non-parametric tests usually have a very bad statistical power, which means you need a lot of data to identify independencies [146]. This problem gets worse the more variables you condition on – conditioning can be viewed as reducing the data you can test with. Incorrect conditional dependencies can lead to incorrect causal conclusions.

Let's say you found a set of statistical independencies, how does this help you to learn the causal graph? Well, statistical independencies in the data are only compatible with certain kinds of causal structures. The data narrows down the causal stories that make sense. So the question is, how many *faithful* causal graphs (see the box: "Central assumptions in causality") are compatible with the data?

As an example, assume you had only three variables: fatigue, COVID, and fever. By running statistical tests, you find that

- fatigue is independent of fever given COVID (Formally, fatigue $\perp\!\!\!\perp$ fever | COVID)
- None of the three variables is unconditionally independent of the other.

Then, three causal stories (causal chains) would explain this data.

- Story 1: fatigue → COVID → fever.
- Story 2: fatigue ← COVID ← fever.
- Story 3: fatigue ← COVID → fever (called fork structure)

All of these stories (or causal graphs) are faithful and compatible with the given statistical (in-)dependencies. Also, one can show that there is no other causal graph that satisfies this.

> **i** The elements of causality: chains, forks, and immoralities
>
> Let's assume you have the variables X, Y, and Z. Three basic path structures in causal graphs are distinguished in causal discovery:

- *Causal chains*, where Y is called a *mediator*:

- *Forks*, where Y is called a *common cause*:

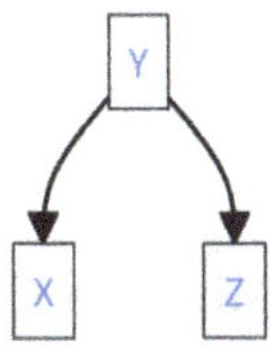

- *Immoralities*, where Y is called a *collider*:

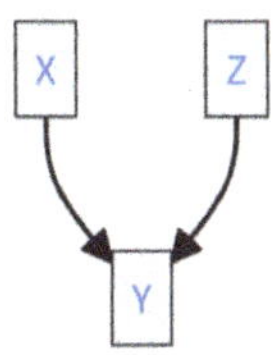

Causal chains and forks imply identical conditional dependencies:

1. $X \perp\!\!\!\perp Z \mid Y$ (X and Z are independent given Y)
2. X and Y are dependent
3. X and Z are dependent
4. Y and Z are dependent

Immoralities, however, exhibit a distinct pattern:

1. $X \perp\!\!\!\perp Z$ (X and Z are marginally independent)
2. X and Z are dependent conditional on Y
3. X and Y are dependent
4. Y and Z are dependent

We call the set of faithful causal models that satisfies the same set of statistical (in-)dependencies the *Markov equivalence class.* How can you find the Markov equivalence class if all you have is a dataset? We provide one of many answers here, the so-called PC algorithm.[6]

The Peter-Clark (PC) algorithm: The most well-known algorithmic approach for finding the Markov equivalence class of causal models that is compatible with the statistical (in-)dependencies found in a dataset [148]. It requires causal sufficiency and the faithfulness condition to be satisfied. It runs the following steps:

1. **Identify the skeleton:** We start with a fully connected network with undirected arrows. Then, step by step we look through all unconditional and conditional independencies between features. We start with unconditional independencies, continue with independencies conditional on one variable, then two, and so on. If two features are (conditionally) independent, we erase the arrows between them. In the end, we get the so-called skeleton graph without directed arrows.

    - *Unconditional independence:* By running a statistical test, we learn that BMI, COVID-vac, flu-vac, and density are all pairwise independent. This allows us to erase the arrows between those variables.
    - *Independence conditioned on one variable:* Conditioning on COVID makes fatigue, fever, and appetite pairwise independent. Moreover, BMI, COVID-vac, flu-vac, and density become independent of fatigue and fever. We can erase all arrows between these variables.
    - *Independence conditioned on two variables:* If we condition on COVID and density, BMI, COVID-vac, and flu-vac become independent of appetite. We can therefore erase all arrows between appetite and these variables.

[6]The PC algorithm is already quite old but great for understanding the general idea of causal discovery. There are many extensions, generalizations, and alternatives of PCs on the market. We point you to [143] for an overview and to [147] for an R package implementing standard causal discovery algorithms.

- We find no further independencies. So this is our resulting skeleton!

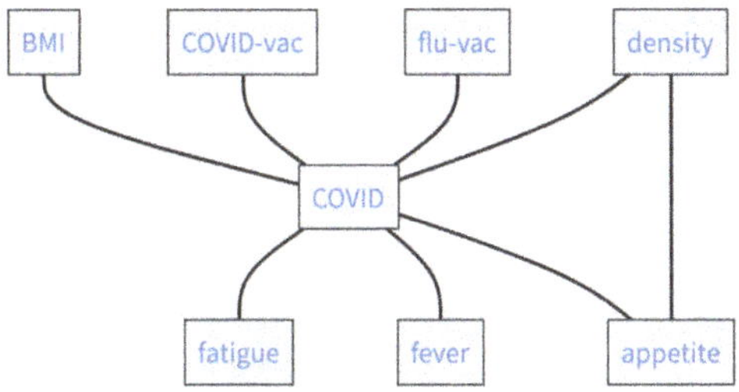

2. **Identify immoralities:** Immoralities are paths of the form $X \rightarrow Y \leftarrow Z$ that lead to unique (in)dependencies, namely to $X \perp\!\!\!\perp Z$ and X, Z are dependent conditional on Y. We can use this uniqueness to orient some arrows. Go through all unconditional independencies $X \perp\!\!\!\perp Z$ where the nodes are connected via one intermediate variable Y, check whether conditioning on Y makes them dependent. If yes, you found an immorality and can orient the arrows.

   - As you can see, in our skeleton all arrows are undirected. We now check for all unconditionally independent variables that become dependent if we condition on one variable. This is the case for BMI, COVID-vac, flu-vac, and density. They are unconditionally independent but become dependent if we condition the COVID variable. We can therefore orient some arrows in our graph.

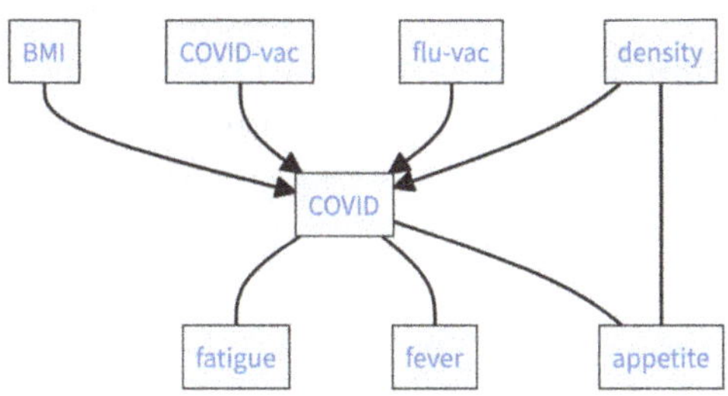

3. **Apply logic:** Now you have identified all immoralities. Thus, if you find three variables that are connected like this $X \to Y - Z$ where the arrow between Y and Z is undirected, you can infer that $X \to Y \to Z$, because otherwise there would be another immorality that you must have discovered. Also, we search for an acyclic graph. So if there is only one way to avoid making a cycle, that is the way you orient the arrows.

    - Which arrows would lead to new immoralities? If fatigue, fever, and appetite had arrows towards COVID, new immoralities would emerge that would have been found in the statistical testing. Thus, they must have arrows coming from the COVID variable. Now only one arrow is left undirected, the arrow between density and appetite. If this arrow would go from density towards appetite, there would be a cycle: COVID → appetite → density → COVID. Thus, the arrow must go from density to appetite.

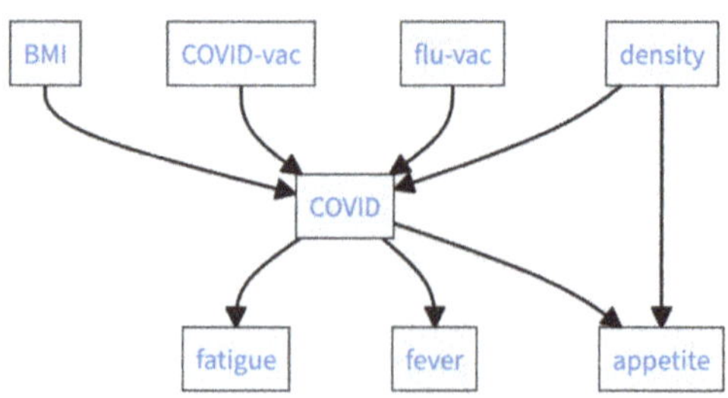

Cool, we were able to identify the full causal graph just from the (in-)dependencies! But we were lucky. If we had looked at a more complex graph, the PC algorithm would have spit out a graph where some arrows are left unoriented. All possible ways for orienting these remaining arrows define the Markov equivalence class. But can you go further? Is there a way to identify the correct causal model among the Markov equivalence class? You need something extra for this:

- **Perform real-world experiments:** You could simply run experiments. There are proven bounds for how many experiments you have to perform to identify the correct causal graph. For instance, if multi-node interventions are allowed, $log_2[n] + 1$ interventions are sufficient to identify the causal graph, where n is the number of nodes [149].
- **Pose more assumptions:** One of the most prominent assumptions for the identifiability of the causal graph is the linearity of the relationship and non-Gaussian noise terms in the structural equations. Alternatively, you can assume non-linearity and the additivity of noise. Check out [143] to learn more. Most approaches rely on the idea that the noise variables are independent in one but not in the other causal direction.

> **i** Central assumptions in causality
>
> You can see causal models as devices for interpreting data. But when are the interpretations correct? Only if certain assumptions are satisfied. Causality is therefore a deeply assumption-driven field. The most crucial ones are listed here:
>
> - **Causal sufficiency**: This is one of the key assumptions in the field. It is sometimes also referenced as the assumption of **no unobserved confounders**. It means that you have not missed a causally relevant variable in your model that would be a cause of two or more variables in the model. This assumption is not testable and without it, far less can be derived [150].
> - **Independence of mechanisms:** Means that causal models are modular. Intervening on one structural equation does not affect other structural equations. Or on a probabilistic take, changing the marginal distribution of one variable does not affect the conditional distributions where this variable acts as a causal parent. This assumption seems almost definitional to mechanistic modeling and implies the independence of the exogenous noise terms [143].

- **Faithfulness:** Is connected to how to interpret causal graphs probabilistically. Faithfulness means that every causal dependence in the graph results in a statistical dependence. Note that it is hard to justify faithfulness in finite data regimes [151].
- **Positivity:** For every group of the population, you have both subgroups with and without treatment. This is crucial when it comes to *treatment effect estimation*. If positivity is violated, you extrapolate (non-)treatment effects for certain subgroups, which can go terribly wrong.

This list is far from complete, there are many more, such as consistency, exchangeability, and no interference [129], [152]. The correctness of your interpretations of data with causal models or the causal models you may partially derive from data are extremely sensitive to these assumptions. If possible, check if they apply!

10.8. Learning causal representations from data

Causal discovery only makes sense if the features are meaningful representations and you can talk meaningfully about their causal relationships. This is the case with highly structured, tabular data that has been heavily pre-processed by the human mind. But the data you often have, especially the data you want to analyze with machine learning, doesn't look like that, it consists of:

- images made of pixels
- texts made of letters
- sounds made of frequencies

For example, it makes no sense to construct a causal model where the variables are single pixels. However, it may be meaningful to talk about the causal relationships between objects in images. How can you go from pixels to these objects? How can you get meaningful higher-order representations from lower-order features?[7]

Machine learning is a field concerned with learning higher-order representations. Causal representation learning therefore describes a dream that many in machine learning share – the dream of fusing symbolic approaches like causal models with modern machine learning. This would combine the strengths of both worlds: learning complex meaningful representations from data AND reasoning symbolically in a transparent and logical manner.

Indeed, doing so is extremely difficult and the research in this field is still in its infancy. Machine learning models learn complex representations to perform their predictions, however, whether the representations are in any sense similar to the ones humans form or at all understandable to us is an open question. Adding labels to the representations you want and modifying the loss function to use these representations might be one way to go [154]. But it is data costly! Is there a way to make sure that machine learning models learn meaningful representations that could be used to construct causal models? Can you maybe even build in constraints informed by your understanding of causality to learn such representations? We'll give you some hunch in this direction...

> **i** Variational autoencoders (VAEs)
>
> Autoencoders consist of two parts. An encoder maps the input to a higher-order representation and a decoder maps the higher-order representation back to the original input space. The only difference in a variational autoencoder is that input is encoded into a set of parameters of a probability distribution, whereas the decoder takes a sample from this distribution and maps it back to the original input space. VAEs are (unlike the rest of the book) unsupervised learning tech-

[7] Causal abstraction is a cool framework for thinking about the relationship between low-level and high-level causal representations [153]. Science is all about causal models at different levels of description, think of the relationship between physical models, chemical models, and biological models.

niques. They allow, among other things, to compress high-dimensional information into low-dimensional features with little or no loss of information.

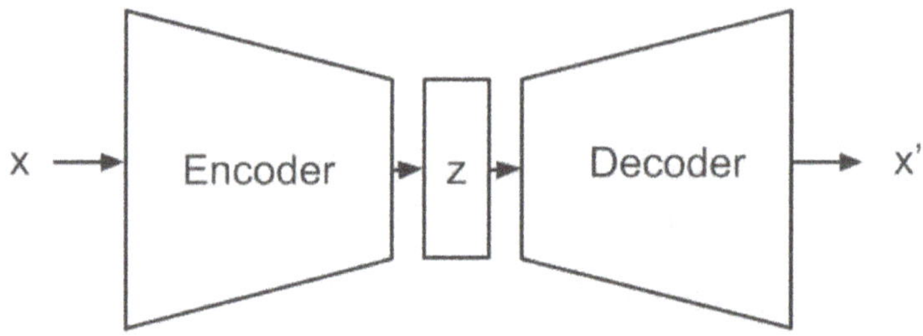

Figure 10.3.: Structure of a Variational Autoencoder.

Putting causal constraints in variational autoencoders: You can see the probability distributions in the middle of VAEs as abstractions of the low-level features your data lives in. Therefore, VAEs are constantly discussed in the context of representation learning in general and there are many ways to incorporate knowledge about these representations both via the architecture and the loss function [155]. Instead of learning representations and then looking at how you can incorporate them in causal models, you can do the reverse and ask: What makes representations good candidates for causal models and how can such knowledge be incorporated into the VAE framework? Since research on the topic is still in its infancy, we will only provide a list of a few ideas inspired by [156] :

- **Sparsity:** One key idea in causality is that a few powerful representations are enough to explain all kinds of complex phenomena. This can for example be enforced by the size of the middle VAE layer.
- **Generality across domains:** Powerful representations are useful across tasks. What makes them useful is that they capture robust patterns in nature. You can enforce this for instance in VAEs by training them to use the same representations for different decoding tasks.

- **Independence:** First, if two representations contain the same information, one of them is redundant. But you want to have simple models to efficiently communicate about the world. Second, the core idea in causality is that you have independent sources of variation and that you can disentangle these sources and their interactions. One approach for achieving this in VAEs is modifying the loss to make sure that the representations in the intermediate layer are statistically independent.
- **Human interventions:** What makes a variable suitable for causal modeling is that you can intervene upon it. Thus, 42 % of randomly selected atoms from a table do not constitute a good representation, as you cannot use or intervene in these 42% in isolation. This human-centered causal bias can be entered into VAEs for example through video data, where objects are constantly moved and intervened upon but stay persistent.
- **Simple relationships:** Causal models are rarely densely connected. Instead, a handful of causal relationships between higher-order variables give rise to the associations you observe in the world. This inductive bias of simplicity can be enforced via the decoder architecture or the loss function.

10.9. Causality helps to formulate problems

After finishing this chapter, you may feel a little overwhelmed. A lot of theoretical concepts were presented and at the same time, there was little practical advice, e.g. compared to the domain-knowledge chapter (see Chapter 8). We have the impression that the link between causal theory and the practical problems of scientists is underdeveloped in current research. For example, there are only a few examples where causal discovery algorithms have been used to gain insights from practice.

Nevertheless, we believe that every researcher should be familiar with the concepts presented above, such as the Reichenbach principle, treatment effect estimation, and causal modeling. Causality is what many of you are ultimately looking for when you want to control, explain, and reason about a phenomenon. While causality as a field does often not provide ready-made practical solutions, it offers you a language to formulate your problem and describe possible solutions.

Many of the chapters in this book are closely related to causality:

- Some domain knowledge (see Chapter 8) you want to encode may be causal.
- Many questions addressed via model interpretation (see Chapter 9), such as algorithmic recourse, are ultimately causal questions [142].
- Causality is one leading approach to improving robustness (see Chapter 11) [157].
- Approaches for better uncertainty quantification, such as conformal prediction (see Chapter 12) are currently being integrated into causal inference [158].

11. Robustness

Machine learning systems should not only work under laboratory conditions – they should work in the wild! And yes, we mean that in the true sense of the word.

Imagine you are an animal ecologist studying the diversity and conservation of species in the Serengeti. You know that machine learning systems allow you to identify, count, and even describe animals from images alone, as illustrated in Figure 11.1. There are tons of motion sensor cameras throughout the Serengeti. Together with the predictions by the machine learning model of Norouzzadeh et al. [7], you'll soon have an amazing dataset to tackle your research questions.

Norouzzadeh et al. [7] have indeed done an impressive job with their machine learning model: Their ensemble of trained convolutional neural networks (CNN) achieves 94.9% accuracy on the Snapshot Serengeti dataset [159] – that is a comparable performance to human labelers. Sounds like a reliable tool to build your research on, right?

To human eyes, Figure 11.2 is the same image as Figure 11.1, but for the model, it is not. The model gets it all wrong now: the species, the count, and even the description. What has happened? The image is a so-called *adversarial example*. All current machine learning models, especially image classifiers, are susceptible to such well-engineered variations in the input that are imperceptible to the human eye but lead the prediction model astray. But Norouzzadeh et al. [7] didn't do anything wrong. Quite the contrary! The paper is a role model for machine learning in science:

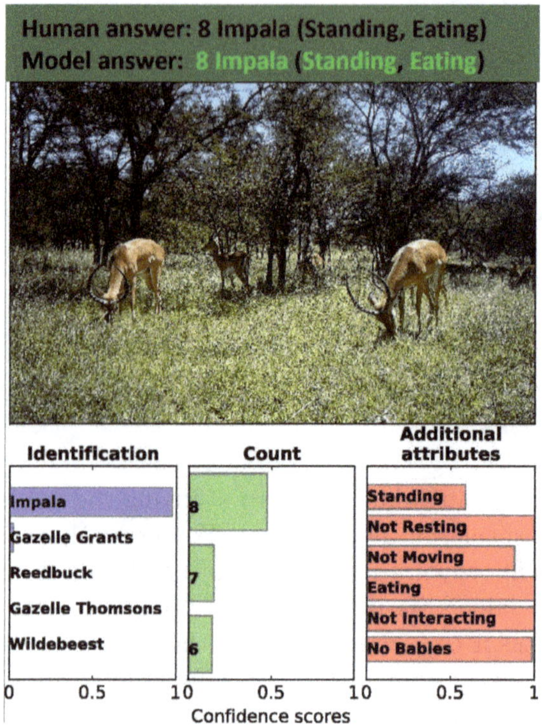

Figure 11.1.: The image was taken by motion sensor cameras in the Serengeti and shows wildlife animals. The machine learning model by [7] correctly identifies, counts, and describes the animals in the image. Used with permission from [7].

- They explain in detail how they deal with class imbalances in the data and target leaks (see Chapter 7), which can occur because the camera sensors take three photos in succession.
- They take into account label noise and conduct confidence thresholding (see Chapter 12).
- They report overall performance and class performance (see Chapter 14) and provide usable open-source code (see Chapter 13).

Still, what should we make of the fact that machine learning models are not robust to adversarial examples?

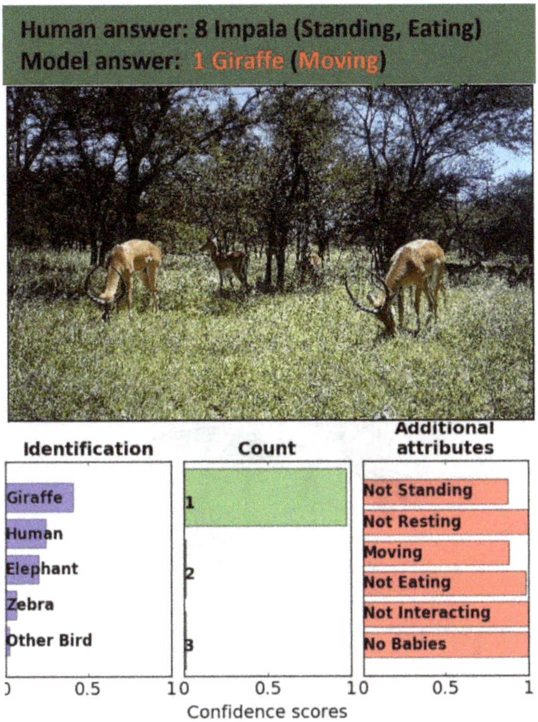

Figure 11.2.: The image looks the same as in Figure 11.1, but now the model is completely wrong in all aspects. Figure modified with permission from [7].

Adversarial examples have kicked off the robustness debate in machine learning [34], and we will explore their occurrence in some detail at the end of this chapter Section 11.7. But not any image would fool the animal classifier, the model performs quite well on real images taken in the wild.

We believe that scientists should not get sleepless nights from deceiving artificial input data that never occurs in practice. In science, there is usually no adversary that deceives your model – except perhaps Reviewer 2. As you will see in this chapter, there are many more profound robustness issues in machine learning that should worry you!

Machine learning not only found its way into science, but its products became global bestsellers. Krarah's former Ph.D. students wanted to get their piece of the pie and turn the tornado prediction idea into a startup. But there was a problem. While the original model worked perfectly in their home territory, it was useless in other countries. They asked Rattle for a workshop to teach them about robust machine learning.

11.1. What does robustness mean?

In everyday language, robustness is a property of a single entity. A washing machine can be robust if it works for many years without trouble. A person can be robust if she can handle many situations competently. However, this leaves open what robustness is really about except for functioning well in general.

To detect robustness problems and solve them systematically, you need a language that allows you to operationalize *robustness* [28]:

- **Robustness target** is the thing that should be robust. For example, you may be interested in how robust the performance of the animal classifier is.
- **Robustness modifier** is the thing with respect to which the robustness target should be robust. For example, this could be the images on which you apply your animal classifier.

- **Modifier domain** specifies the relevant changes in the modifier to which the target should be robust. For example, this could be changes in the background lighting of the images.
- **Target tolerance** specifies how much the target is allowed to change if the modifier changes within the modifier domain. For example, you might be fine if the model performance decreases a bit for images with a darker background as long as the performance does not drop drastically.

So what does robustness mean?

> **Definition:** A robustness target is robust with respect to a modifier if relevant interventions to the modifier within the domain do not lead to greater changes than specified by the target tolerance.

We can now talk in a more nuanced way about robustness. For example, the wildlife image classifier performance (robustness target) is relatively robust to changes in the lighting conditions when taking the images (relevant interventions). However, it is less robust to targeted modifications (irrelevant interventions) on the input pixels. Ultimately, to judge whether your model is suitable for a specific application, you have to check if it is robust to relevant interventions on modifiers. This forces you to think about the changes you expect to occur in deployment!

11.2. Auditing and robustifying strategies

Broadly speaking, robustness researchers have to adopt two different perspectives:

- **Auditing:** Is your model performance robust to relevant modifier changes? For example, you can check how the animal classifier performs when you darken the background of the test data.
- **Robustifying strategies:** What can you change in data collection and model selection to make your model more robust to relevant modifier changes? For example, if you train

your model with more images taken at night, your model will generally perform better in this environment.

Both aspects are interacting. You may audit the model and detect potential robustness weaknesses. To mitigate these weaknesses, you apply robustifying strategies. But what should you audit for?

11.3. Understand data distribution shifts

In practice, the most crucial modifier is the data. Your training data may stem from one source, but when you now deploy your model, the data may look entirely different. There are different *sources* of data distribution shifts:

- **Natural shifts** occur because the natural conditions change. If you think of the animal classifier, you have to deal with:
  - varying weather or lighting conditions,
  - changes in the flora and fauna over time, or
  - new camera sensors placed at novel locations.
- **Performative shifts** are induced by the model itself and its effects on the data. Imagine a model that predicts which diseases a person will develop based on their behavior. Provided with these predictions, the person may change her behavior and thus invalidate the prediction.
- **Adversarial shifts** occur due to attackers who modify the data. An example is given in Figure 11.2.

For natural scientists, the most important distribution shifts are natural distribution shifts, whereas social scientists must care just as much about performative shifts. Adversarial shifts are more of a problem in business or industry applications but less so in science.

You can also distinguish different *types* of data distribution shifts. Data in supervised machine learning is described as pairs $(x^{(i)}, y^{(i)})_{i=1,\dots,n}$ sampled from an underlying distribution $\mathbb{P}(X, Y)$, where $X := (X_1, \dots, X_p)$ describes the input features and Y

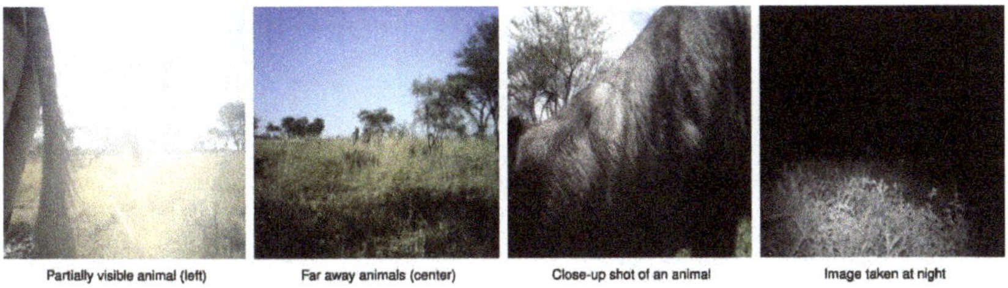

Figure 11.3.: The example shows different natural conditions under which the model is expected to perform well, such as pictures taken from afar, close, or at night. Used with permission from [7].

the target variable. We can express the different types of data distribution shifts directly through the distributions of X and Y:

- **Covariate shift** describes a case where the distribution $\mathbb{P}(X)$ has changed. In the example above, a covariate shift can occur if a camera was previously in the open sun but is now in the shade of a plant.
- **Label shift** describes a case where the distribution $\mathbb{P}(Y)$ has changed. The installation of a new camera in the previously ignored riverine forests in the Serengeti, for example, will lead to many more hippos being observed.
- **Concept shift** describes a case where the conditional distribution $\mathbb{P}(Y \mid X)$ has changed. In the example of wildlife, this can happen when there is a new categorization of species; previously there is only the gazelle category, but afterward it is subdivided into Grant's gazelle, Thomson's gazelle, and so on.

Again, understanding what type of distribution shift you face can be vital for robustifying your model. But before we come to robustifying strategies, let's first look into robustness audits.

11.4. Strategies to audit for robustness

Has the distribution shift already occurred? Or are you only anticipating a distribution shift? This determines which audit you can perform:

- **Post-hoc audit:** The distribution shift has occurred already. You have collected new data after the shift and can evaluate the quality of your model on this data. In the wildlife example, you may have only trained your model with data from the dry season, and now, in the middle of the wet season, you want to evaluate the model's performance.
- **Anticipatory audit:** The distribution shift has not yet occurred. So you have no data about the expected changes. For example, if you are in the dry season and want to predict how your model will work in the wet season.

11.4.1. Post-hoc audit

A post-hoc audit is comparably simple – just analyze your model performance on your data. For example, you might want to compare the model performance between dry (pre-shift) and wet season data (post-shift).[1]

Understand whether and how the data differs

But how can you even recognize that the distribution has shifted? You need to constantly monitor your data and check the data properties. For tabular data, there are a variety of summary statistics like the mean values, standard deviations, ranges, or correlation coefficients. The situation is similar for text, where you have word frequencies, word lengths, similarity of word embeddings, cosine similarity, or text sentiment. When these statistics start to vary, your distribution shift alarm bells should ring. For image data, we lack good summary statistics. Instead, eyeballing data samples

[1]To evaluate your model performance on the new data it must be labeled. We discuss in Section 11.5.3 how to label data systematically using active learning.

through time taking into account domain knowledge can be more effective.

There are also modality-independent automated strategies for detecting data distribution shifts – called *out-of-distribution (OOD) detection* [160]. We discuss them below in Section 11.5.2.

Analyze the different errors

Just as important as understanding the distribution shift itself is understanding how it affects model performance. Comparing performance before and after the shift is only the tip of the iceberg of a proper audit. False positive or false negative rates may also differ, and you could have contextual preferences for either bounding false positives or false negatives. Similarly, we recommend to group and compare errors across output classes.

Interpretation methods can point to the sources of errors

Interpretation methods (see Chapter 9) give you insight into the features on which the model relies upon and how these features impact model performance. Feature attribution techniques, for example, allow you to analyze the attention of image classifiers for specific predictions. You should be alerted if the model classifies Grant Gazelles by looking at the background rather than the animals [99]. If you deal with tabular data, you may compare global feature importances based on data before against after the distribution shift.

11.4.2. Anticipatory audit

An anticipatory robustness audit is more demanding than a post-hoc audit. The shift has not occurred yet – so you have no data from after the shift. You need to specify the shift qualitatively based on the sources and types of shifts above. To also quantify the effect of an anticipated shift on model performance, you additionally have to generate synthetic data that reflects the shift.

Specify the shift qualitatively

What shift do you expect, a natural or a performative shift? Is it a shift in covariates, in labels, or even in concepts? Be careful to specify the shift correctly. For example, auditing your model for adversarial robustness in a non-adversarial setting is pointless. Consult your domain knowledge! What aspects of your domain do you expect to vary and to which of the above types of shifts do they belong? Think back to our wildlife example: researchers are aware that there are relevant seasonal effects on the Serengeti environment and its wild inhabitants.

Generate (semi-)synthetic data and systematically test your model on it

To perform a quantitative audit, you need data that reflects the distribution shift. But because you don't observe but only anticipate the shift, you have to generate data synthetically. You have to translate your qualitative knowledge about the shift into a way to generate data. We discuss various strategies to generate data below in Section 11.5.3, for example, image filters allow you to turn day data into night data (see Figure 11.6).

Suppose you have created a synthetic dataset, does this mean you finally reached the point you start at in a post-hoc audit? Yes and no, there remain two big differences: 1. You already have a profound understanding of the data because you generated it. 2. The insights your synthetic data offer depend on your assumptions about the shift.

Interpretation methods allow you to anticipate shifts

Interpretation methods allow you to perform an audit without specifying the expected distribution shift or synthetic data. Instead, you simply analyze your pre-distribution shift data in a more explorative way, e.g. if feature attribution methods indicate that the model relies on background features like trees to classify an animal, this could mean that the model performs worse under varying background conditions. This alerts you to both the potentially dangerous distribution shift (changes in background) and the reason why your model is failing (reliance on background). Similarly, if a spurious feature has a large feature importance, your model may fail in settings where this feature varies; A COVID risk predictor trained in Munich that relies on street names is likely to fail in Paris. Finally, feature effect methods such as individual conditional effect curves (ICE) and counterfactual explanations describe model behavior in relevant counterfactual scenarios.

> **i** Context matters in robustness audits
>
> What you need to audit your model for, depends heavily on the application context. The wildlife model, which acts as a labeling tool, requires less extensive auditing than machine learning-based medical diagnosis. What audits are needed depends on the risks of errors, the instability of the environment, and other domain characteristics. For high-risk applications like in medicine, there are often additional legal requirements, such as those set out in European AI legislation and in medical device regulations.

11.5. Strategies to robustify your model

Say your audit has shown that your model is not robust to relevant distribution shifts. How can you robustify it? This is a list of common robustifying strategies, ordered by their position in the machine learning pipeline. Each of the strategies is discussed in more detail below:

1. **Control the source:** Some shifts are under the control of the model authority. For example, the purchase of a new brand of camera sensor can lead to greater distribution shifts than sticking with the old brand.
2. **Filter out-of-distribution data:** You could train a second model to filter out data that significantly differs from the training data. For example, the machine learning model should not provide predictions for animals that are not included in the training set, leaving these cases for human labelers.

3. **Gather data representative of the shift:** You may gather additional (real or synthetic) data that accounts for the shift. For instance, one could use image filters to augment wet season data using dry season data.
4. **Carefully select and construct features:** The lack of robustness in your model may be due to its reliance on incorrect features or improper feature encoding. For instance, by removing the image background, shifts in the background will no longer affect the model's performance.
5. **Choose inductive biases wisely:** Some shifts can be accounted for by adjusting your modeling assumptions (see Chapter 8). For counting animals, you may choose an architecture that generalizes to high animal counts like sequential subtizing [161].
6. **Transfer learning to obtain robust representations:** Often, your model is sensitive to distribution shifts because its learned representations overfit the training data. In such cases, *transfer learning* – reusing representations learned by other models trained on the same data modality – can help. Norouzzadeh et al. [7] demonstrate that transferring representations from common image classifiers increases the data efficiency of the wildlife classifier.

The strategies you should choose to robustify your model should be informed by your audit. If you caused the distribution shift yourself, you might be able to control the shift. When your model performs poorly only on rare anomalies, filtering out these cases might be the best approach. If the distribution shift is unavoidable and causes a significant performance drop, the most common strategy is to augment your data and retrain the model. If the distribution shift is unavoidable but data augmentation is difficult, you may need to engineer features, adjust the modeling assumptions, or use transfer learning.

11.5.1. Control the source

It is usually not like distribution shifts are just happening and there is nothing you can do about them. You often have an active role – you caused the distribution shift by an action. You installed a camera sensor from a different brand and suddenly the performance drops? You put food next to the camera to spot more animals and suddenly your animal count bound reaches its limits? In some cases, rather than adapting the data or the model, it can be smarter to tackle the very source of the shift. You must enact control to guarantee a stable environment, e.g. by using only cameras from the same brand.

But not all distribution shifts are within your control. You cannot keep the Serengeti in the wet season all year. Other sources you can control but not without unwanted side effects: You can create stable lighting conditions around the camera, but this can attract or repel certain animals.

11.5.2. Filter out-of-distribution data

There will always be data that presents a challenge for your model. Especially data that is very different from the training data. Out-of-distribution (OOD) detectors enable you to filter out data on which your model would perform substantially worse. These data can then be handled separately, e.g. by a human labeler. Thereby, OOD detection robustifies overall performance in deployment. However, OOD detectors only act as filters; they do not help with substantial distribution shifts.

How to find out if a data point is OOD? With OOD detectors! Sometimes people further differentiate between methods designed to detect *anomalies* (rare data from different distributions), *novelties* (data from a shifting distribution), or *outliers* (rare data within training distribution) [160], [162]. Here, we focus on OOD detection more generally. There are four different approaches [160]:

- **Classification-based methods** phrase OOD detection as a classification problem. Some classification-based methods require data labeled as within and outside of the distribution. Others leverage uncertainty-aware classifiers and label data with high classification uncertainty as OOD.
- **Density-based methods** model explicitly the probability density of the training data. Data whose density lies below a certain threshold is labeled as OOD.
- **Distance-based methods** calculate the difference of a given data point to the centroid or prototypes in a dataset. Data whose distance surpasses a certain threshold is labeled OOD.
- **Reconstruction-based methods** use autoencoder methods to detect OOD data. Data with higher reconstruction error are labeled as OOD.

To get an overview of all the different types of OOD detectors check out the review papers by [162], [160], and [163].

Reconstruction-based techniques have advantages over the other approaches: Unlike classification-based methods, you do not need to label data as (non) OOD; Unlike density-based methods, you do not need to specify a model of the probability density; And unlike distance-based methods, you do not need to construct complex reference points such as centroids or prototypes. Let's therefore take a deeper look into reconstruction-based OOD detectors.

> **i** Autoencoders
>
> Autoencoders describe a neural-network-based method to project high-dimensional inputs into a low-dimensional feature space (often called *latent space*) with a minimal loss of information. This is achieved through a two-stage architecture:
>
> 1. **Encoder:** Projects the high-dimensional input into a prespecified low-dimensional latent space.
> 2. **Decoder:** Maps the projected inputs from the latent space back to the original space.

The encoder and decoder mappings are optimized to minimize the reconstruction error of the training data. The reconstruction error of a given input describes the difference (according to some metric) between this input and the output we obtain after sequentially encoding and decoding the input. In our example, this would mean calculating the mean squared error between the initial Impala image and its reconstructed version that runs through the encoder and decoder network.

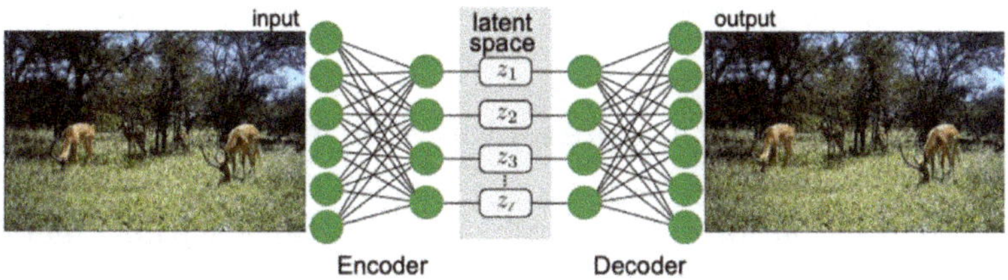

Figure 11.4.: Structure of an autoencoder. Figure created by combining elements from [164] and [7] (used with permission). All rights reserved.

Out-of-distribution detection with autoencoders

Say x is the data point you want to classify as within or outside of the training distribution. Then the encoder network can be described as a mapping $E : \mathbb{R}^h \to \mathbb{R}^l$ from a high-dimensional feature space $\mathbb{R}^h$ to a low dimensional latent space $\mathbb{R}^l$. Similarly, the decoder network is a mapping $D : \mathbb{R}^l \to \mathbb{R}^h$ and L is a distance function on the input space $\mathbb{R}^h$ (e.g. mean squared error). Then, the *reconstruction error* of x is defined as:

$$\mathrm{RE}(x) := L(x, E(D(x))).$$

This allows us to define a simple OOD detector:

$$\mathrm{OOD}(x) := \begin{cases} \text{OOD} & \text{if } \mathrm{RE}(x) > \tau \\ \text{not OOD} & \text{else} \end{cases}$$

Intuitively, any data with a reconstruction error above a certain threshold τ counts as OOD. This primitive approach faces clear limitations:

- Also OOD data can have low reconstruction error. By extending the reconstruction error with the Mahalanobis Distance on the latent space, this problem can be tackled [165].
- Training autoencoders is hard, you have to make architectural choices based on domain knowledge (e.g. CNN for images), define an appropriate latent space (i.e. just the right size to capture the training distribution without information loss), and choose an appropriate loss function [162].
- Training autoencoders requires representative training data [163], but in practice, the data is often unbalanced.

11.5.3. Gather data representative of the shift

Why are distribution shifts a problem for machine learning models? In the case of a covariate shift (i.e. shift of $\mathbb{P}(X)$) or a label shift (i.e. shift of $\mathbb{P}(Y)$), your model fails because it has never seen this kind of data and is unable to extrapolate. In the case of a concept shift (i.e. shift in $\mathbb{P}(Y \mid X)$), your model may have faced similar inputs but the learned dependencies became unreliable.

In any case, the most prominent solution in the research literature for how to robustify your model to such shifts is the same – gather more data that reflects the distribution shift! There are two strategies to obtain such data:

- **Gather real labeled data with active learning:** You may want real labeled data. Active learning concerns the systematic search for data worth labeling.
- **Augment your data:** Gathering real data comes with high costs and you have limited control over which kind of data you get. Data augmentation is concerned with generating synthetic instances using domain knowledge.

Active learning

Which data should you label to robustify your model? Labeling all data is often too expensive and time-consuming. The literature differentiates three active learning setups, based on how the labeler receives the data [166]:

1. In *membership query synthesis* any input in the input space is a potential candidate for labeling, even completely unrealistic inputs.
2. In *stream-based sampling* you receive data sequentially and have to decide whether to label this data.
3. In *pool-based sampling* you receive a big data sample and have to choose a subsample that you want to label.

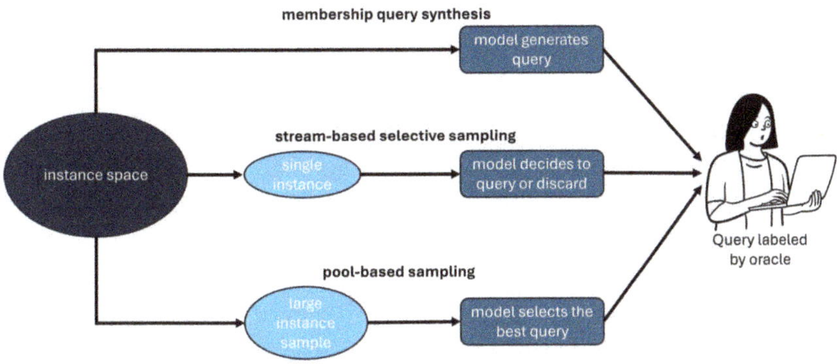

Figure 11.5.: The three main active learning setups. Based on concepts presented in [166].

Active learning is largely concerned with the automation of the selection process for labeling [166], [167]. The data selection process can be based on high prediction uncertainty, proximity to the decision boundary, randomness, expected model error, expected training effect, or data representativeness. In the example, Norouzzadeh et al. [7] suggest labeling those wildlife images with the highest prediction uncertainty. Note that all these different active learning strategies are readily available in recent Python packages like modAL [168] and ALiPy [169].

You may also have intuitions about what data to track and label. Incorporating such knowledge [170] and interacting with the model through interpretability techniques [171] can significantly improve active learning strategies.

Data augmentation

Data augmentation is about creating synthetic data. Wait, don't you also need active learning to label these data? In some cases yes, namely if you want to find the label for an arbitrary input. However, when we talk about data augmentation, we usually create data for which we know the label. Data augmentation has been particularly the focus in computer vision [172], [173], but recently there is growing literature on data augmentation in the context of natural language processing [174].

There are two general strategies to augment your data [172]:

- **Data transformation:** Transformations applied to labeled data that are known not to change the label. Selecting the right transformations is an excellent way to incorporate domain knowledge. Focus on transformations you expect to occur in practice.
- **Data synthesis:** Creation of entirely new data with known labels. Data synthesis may be based on generative models or Computer Aided Design (CAD) models. The aim is to generate synthetic data that shares important properties with your training data but varies from it in relevant aspects.

In computer vision, there are geometric transformations (changing the angle, position, direction, or size of an image) and photometric transformations (changing attributes like coloring, saturation, contrast, or camera artifacts). Some transformations concern the overall image, while others only concern one region. Transformations may delete, replace, swap, or recombine regions of the image.

In natural language tasks, it is common to randomly insert, delete, or swap words in a sentence or to substitute words in sentences with their synonyms. If you have a trained language model, also more sophisticated transformations can be performed, such as: back translation, where sentences are translated back and forth between two languages; text paraphrasing, where the same content is rephrased without changing the meaning; and style transformations, where the same content is described in different styles (e.g. more formal vs less formal language).

Figure 11.6.: Impala image (left) with different relevant transformations applied: simple rotation, background from day to night, and artsy filter. Simple rotations can be implemented by hand, whereas complex filters often rely on generative models. Figure modified with permission from [7].

Data synthesis is more challenging. Computer-aided design (CAD) allows you to model physical objects by hand and perform various geometric and photometric transformations in a physically adequate way. Neural rendering learns 3D scene representations from 2D images and thereby avoids hand-crafted CAD modeling. The most common approach to data synthesis in computer vision is using generative models like generative adversarial networks (GANs) or variational autoencoders (VAEs). The models generate realistic images that can be sampled conditionally on a desired target class. In natural language processing, the best generator models are large language models like ChatGPT or LLaMA, which can be prompted to generate data with a certain content.

i Generative adversarial network (GAN)

A GAN is a generative model that allows the generation of highly realistic data [175]. It is trained using two sub-models:

- **Generator:** This neural network model is designed to generate data. It takes random noise from a (Gaussian) distribution as inputs and ideally transforms the noise into realistic data.
- **Discriminator:** This network is designed to distinguish real from artificial data. It takes inputs and decides whether they are real data or synthetic data.

The two sub-models are trained iteratively in a zero-sum game against each other. The generator aims to generate data that the discriminator mistakes for real data. The discriminator aims at perfectly separating real from synthetic data. In the course of the training, both models get better and better at their tasks, which ultimately leads to a well-performing generator model.

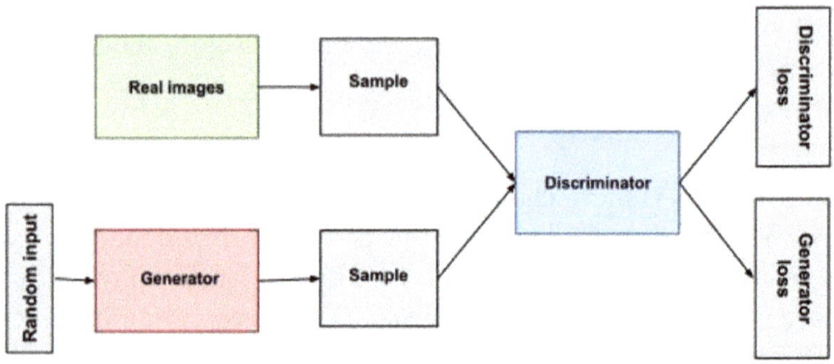

Figure 11.7.: Structure of a GAN. Figure from Google[a], CC-BY (https://creativecommons.org/licenses/by/4.0/)

[a]https://developers.google.com/machine-learning/gan/gan_structure?hl=de

The divide between data transformations and synthesis is less straightforward than what we suggested above. Relevant transformations often go beyond simple geometric or photometric transformations. Think of changing the entire background or adding other animals to the Serengeti images. Complex transformations often demand data synthesis methods. For example, data augmentation GANs [176] and conditional GANs [177] allow the generation of new instances conditioned on a given data instance.

Common packages to perform data augmentation in Python are the Augmentor package for computer vision [178], ImageDataGenerator in Keras, and the natural language toolkit (NLTK) for natural language processing [179].

What happens after you obtain the data?

Let's say you have gathered the required data. What should you do? You could *retrain* the entire model: merge the newly collected data with your training data and run again your machine learning algorithm. Another strategy is to *fine-tune* your existing machine learning model by training it on the new data you collected. Indeed, fine-tuning sounds like less work than retraining but dependent on the fine-tuning specifics it often introduces a bias either towards the training data or the new data. Instead of retraining and fine-tuning, you may decide to train a new *separate model* exclusively on the newly collected data.

If you face a covariate or a label shift, retraining and fine-tuning are both reasonable strategies. Old and new data can be merged in one model as the predictive pattern between the two stays intact. Fine-tuning is particularly advisable if you gain a few high-quality data with active learning and you want to emphasize this data in training. In a concept shift, on the other hand, the predictive pattern between old and new data differs – training a separate model is the only option.

Does data augmentation really improve robustness?

This question is difficult to assess in general as it depends on the domain and the data augmentation approach. It is a mantra in machine learning that more data is always better. But what if the data is synthetic? In many settings, data augmentation indeed effectively improves robustness [172], [180]. However, in some cases, (adversarial) robustness conflicts with predictive performance on the original dataset [181]. This is unsurprising: robustness means insensitivity to certain features, and insensitivity to predictive features leads to a performance drop. Whether you have to trade off (adversarial) robustness and performance depends on the data augmentation approach [182]. Interestingly, training your model exclusively on synthetic data may turn your model mad [183].

11.5.4. Carefully select and construct features

As we discussed in Chapter 2, the input features and the target feature are key modeling choices that every modeler faces. Achieving a robust model with unreliable features can be impossible. Imagine having to predict an animal's species only on the background against which it was sighted. The slightest distribution shift will diminish model performance. Similarly, if you are forced to predict a specific species but 75% of the data contains no animals [7], your model is doomed to fail from the start!

Choosing reasonable input and target features is challenging and requires (causal) domain knowledge (see Chapter 8 and Chapter 10). There exist various approaches to obtain better input features to robustify your models:

- **Feature selection** describes approaches to select an optimal subset of features from a feature set [184]. The subset is chosen based on criteria like the (conditional) mutual information between input and target, or the performance of a trained classifier on the subset of input features (see e.g. conditional feature importance in Chapter 9). Feature selection reduces the dimensionality of the data and filters unreliable or noisy features, thereby improving robustness. Feature selection algorithms are often tailored to tabular data but are less useful for image, text, or speech data.
- **Feature engineering** describes approaches that transform the input features. One option is to apply hand-crafted transformations, which:
    - describe statistical properties, e.g. interaction terms,
    - encode domain knowledge, e.g. graph structure in graph neural networks,
    - have physical meaning, e.g. edge detectors,
    - reduce dimensionality, e.g. linear regression with L1 loss,
    - provide useful encodings, e.g. bucketing or one-hot-encoding, or
    - emphasize particularly important features, e.g. time.

Alternatively one can *learn* transformations that lead to features with desired properties, for example, smoothness, sparsity, or task generality. These so-called *representation learning* approaches often lead to increased robustness under distribution shifts [155].

Also, the target encoding can be improved. Norouzzadeh et al. [7], for example, dissect their prediction target into two parts: In Task 1, a model discriminates between inputs with and without animals; In Task 2, a second model classifies the images that contain animals into different species. This dissection in two tasks substantially improves the classifier's robustness. Similarly, target features often come in a hierarchical form [185]: the Thompson Gazelle and the Grant Gazelle are different subspecies of Gazelles. Such additional structure can be encoded into hierarchical multi-label encodings like decision trees and thereby improve the robustness of the classifier.

11.5.5. Choose inductive biases wisely

The relationship between modeling choices and predictive performance has already been discussed extensively in Chapter 7 and Chapter 8. The key insight was – the better suited the inductive bias (e.g. model class, architecture, hyperparameters, loss, etc.), the faster you will learn a high-performing model. Data distribution shifts are nothing but learning environments. Modeling choices, therefore, provide another robustifying strategy. For example:

- CNNs are translational invariant. The position of the animal in the image does not affect model performance.
- Dropout improves the robustness to adversarial attacks by switching off certain neurons in training [186]. The reason is that dropout enforces smoother representations in the latent space.

Improving robustness to distribution shifts with data augmentation and with inductive modeling biases are two sides of the same coin. The former turns knowledge about (anticipated) distribution shifts into data instances; The latter turns knowledge about (anticipated) distribution shifts into modeling assumptions. CNNs are only one example where this parallelism is evident, many similar architectural solutions encode rotation and scaling invariance [187]. Similarly, graph neural networks allow encoding all forms of domain knowledge into the graph structure and node/edge properties [188], [189], [190].

Is it better to apply data transformations or to encode invariances into the inductive modeling assumptions? On the one hand side, some data transformations are difficult to encode as inductive biases into the model. Think about dry versus wet season data, there is no easy way to encode an invariance to seasonal changes. On the other hand, if it is possible to encode invariances as inductive biases, you should. Your model is guaranteed to obey them, whereas data augmentation only makes it more likely that the model learns invariances. Furthermore, there is a computational trade-off: more data requires more compute, whereas better inductive biases usually improve computational efficiency.

11.5.6. Transfer learning to obtain robust representations

Transfer learning is about transferring knowledge from one task to another. So learning a task does not have to be done from scratch but can build upon existing knowledge. While tasks often differ substantially, they share certain aspects. For example, classifying pets and wildlife animals both require learning higher-order representations that allow us to differentiate animals, even though the animals, their actions, and image backgrounds differ. By inducing more domain-general knowledge, transfer learning makes the model more robust to common distribution shifts. We distinguish different kinds of transfer learning according to the knowledge they transfer:

- **Feature extraction** builds on the representations learned in one task to reuse them in another task. For example, say you have a general image classifier like ResNet or Inception

V3 that has been trained on a huge dataset like ImageNET. Then, there are representations stored in the neural network's weights, activations, and nodes, which can be reused to make the wildlife classifier robust against common image permutations. Commonly used feature extraction methods focus on the penultimate layer before the final prediction. Feature extraction is the most popular transfer learning technique and has also been used by Norouzzadeh et al. [7] leading to better data efficiency.

- **Fine-tuning** uses the learned specifics of a model trained for one task as the starting point for another task. One may, for example, take the trained ResNet classifier, substitute the output layer, and train the model on the wildlife images. Unlike in feature extraction, the model can adapt the representations learned by ResNet on ImageNet and tailor them for the wildlife case. This was again performed by [7], however, without improving overall performance significantly.
- **Multi-task learning** concerns training a single core model that is used to perform multiple related tasks simultaneously. The idea is to enforce the core model to learn representations that are useful across different tasks, leading to improved representations and better robustness to common changes. One could, for example, use the same core model to classify trees, and wildlife animals, and predict daytime. By optimizing for such a diverse set of tasks, the core model has to learn representations that work on all of them.
- **Self-supervised learning** can be seen as one specific approach to learning a core model that is useful across a variety of tasks. It masks certain parts of the data and tries to infer them from the rest of the data. Thereby, self-supervision learns generalizing pattern. One could, for example, mask the animal's heads to learn interdependencies between animal heads and their bodies.

Transfer learning can be key if data is scarce for the specific task at hand but widely available for related tasks. More and more there are core models for all data modalities like ChatGPT for text data or ResNet for image data. Based on vast amounts of data, these core models have learned such powerful representations that they make the models robust to a wide range of standard distribution shifts. We believe that in future scientific applications, transfer learning from core models may play an essential role. The amount of data and knowledge that entered these models should not go unused.

11.6. Generalization and causality are linked to robustness

Remember the three different kinds of generalization that we discussed in Chapter 7? Generalization to predict in theory, generalization to predict in practice, and generalization to the phenomenon. Generalization to predict in theory concerns prediction for a static data distribution, indeed a natural requirement on machine learning models. Robustness usually goes one step further: the machine learning model should perform well in all practically relevant scenarios. That means the performance of the model should be robust under expected data distribution shifts. Generalization to predict in practice and robustness therefore often means the same thing – the model should work under natural conditions.

Causality is about generalization to the phenomenon at hand. You want an accurate representation of the data-generating process. This constitutes a link between robustness to causality. Why? If you have an accurate representation of the phenomenon, then you can simulate all kinds of alternative scenarios and provide predictions under all kinds of distribution shifts:

- **You know what led to the distribution shift?** Then you can simulate the shift using your causal model and still provide optimal predictions with your causal model [191], [192]. Even your initial machine learning prediction model may perform robustly under certain shifts [142].
- **You do not know what led to the distribution shift?** Say you only receive data indicative of the shift. Then, causal

models allow to generate data for various possible shifts and compare them to the observed data [193].

Even beyond these cases, causal models have an essential property that makes them appealing for machine learning robustness research – causal models are modular. Say you train a machine learning model to learn the joint distribution of two variables A and B, namely $\mathbb{P}(A, B)$. Then, as soon as your distribution of A or of B changes, you have to learn an entirely new model.

Instead, assume you know that A causes B. Then you can split your learning task into two components, namely $\mathbb{P}(B \mid A)$ and $\mathbb{P}(A)$. This provides you again with the joint distribution because

$$\mathbb{P}(A, B) = \mathbb{P}(A)\mathbb{P}(B \mid A).$$

But now if $\mathbb{P}(A)$ shifts, all you need to update is your model of $\mathbb{P}(A)$, whereas $\mathbb{P}(B \mid A)$ must remain stable because it is a causal mechanistic relationship (see Chapter 10). This modularity makes causal relations worth learning!

11.7. The riddle of adversarial examples

Why does a model that is as good as in [7] make mistakes like in Figure 11.2? Behind this question lies the riddle of adversarial examples. Adversarial examples are inputs that are modified in a way that makes them humanly indistinguishable from the original input but entirely shifts the model's prediction, leading to misclassification. The reasons for this behavior are still only partially understood.

The first hypothesis was that adversarial examples describe unlikely instances that are not well represented in the original training data [34]. However, if this were true, adding adversarials to the training data should get rid of the problem – but it didn't [194]. Even if you train your model on a wide variety of adversarial examples there will still be new adversarial examples in the direct vicinity. Famously, Goodfellow et al. [195] came up with a new proposal.

They suggest that adversarials arise in machine learning models because the models are too linear. Most machine learning models are still based on relatively linear activation functions like Rectified Linear Units (ReLUs). Thus, changing the input up to a certain norm leads (due to linearity) to a significant change in the prediction. This insight led to a new efficient algorithm to compute adversarial examples, the Fast Gradient Sign Method [195]. Having linear activation functions turned out to be neither necessary nor sufficient for adversarials [196]. New adversarial examples showed up not only for ReLU-based neural network models but for all kinds of machine learning models [197]. Moreover, these examples transfer between different models trained on the same dataset – you could generate an adversarial example for a ResNet model and apply it successfully to another model with an entirely different architecture [198]. Linearity couldn't explain this strange behavior.

Adversarial examples are not bugs, they are features

Ilyas et al. [199] gave a novel explanation of adversarial examples – adversarial examples arise because human perception and machine learning perception operate differently. Machine learning models learn patterns that are stable across the dataset including patterns on which humans don't rely. Like classifying an elephant based on the texture of its skin rather than its shape or its trunk. Slight changes in the elephant's skin will not change a human's classification but it changes the one by the machine.

Ilyas et al. [199] came to this conclusion based on the following experiment: Say you have a trained model and then generate a set of adversarial examples for this model. Now, train a new model from scratch exclusively on these adversarial data (with the wrong labels). Surprisingly, this model performs strongly on the original data with correct labels. The only explanation for this behavior is that the adversarial examples are strange in a meaningful way, adversarials differ in features that allow to classify ordinary data points. This also explains why adversarial examples can transfer between different models trained on the same data because they differ from real data in ways that are usually indicative of a different class. Unfortunately, this explanation of adversarials (which seems plausible) does not come with easy fixes. It implies that adversarials can only be evaded by relying on the very same features as humans. However, doing so would limit the space of possible solutions and thereby the predictive performance. In science, new predictive patterns undiscovered by humans might be what you are ultimately after.

Another factor leading to adversarial examples is that machine learning models cannot distinguish causes from effects or spurious correlations. Thus, changing a feature associated with a different target class can flip the prediction without changing the underlying target [200]. Like changing the elephant's skin texture similar to a hippo's without touching causal features like their shape or their trunk. Adding causal assumptions may therefore protect against certain adversarial examples [157]. But again, this is not an easy fix to use in practice. Adversarial examples remain a phenomenon that you have to live with and it is unclear if getting rid of the phenomenon is always desirable. In science, you might even strive for the weird predictive features invisible to humans.

11.8. Robustness is a constant challenge

Robustness in machine learning is not something you achieve. Robustness is like the boulder that Sisyphus[2] keeps pushing to the top of the hill just to see it rolling down again. But if you want a reliable machine learning model, the only way is to keep on pushing! Your environment will change making your model performance drop eventually. It's important to stay on your watch and constantly adapt.

[2]Sisyphus was a figure in Greek mythology, known for his cunning and deceitful nature. He was the King of Corinth and was infamous for his trickery, which often got him into trouble with the gods. One of the most famous stories involving Sisyphus is his punishment in the afterlife. According to Greek mythology, after his death, Sisyphus was condemned by the gods to roll a boulder uphill for eternity, only for it to roll back down every time he neared the top. This endless and futile task became known as *Sisyphean* and is often used as a metaphor for a task that is never-ending and ultimately feels pointless.

The chapter should have made clear that robustifying models is multifaceted. Each step in your pipeline can contribute and each step should be kept dynamic to be able to react quickly to changes. Each chapter in this book is an ally on your neverending machine learning robustness journey, you need models that generalize (Chapter 7), you have to incorporate domain knowledge (Chapter 8), you must audit your model with interpretability techniques (Chapter 9), you should be aware of prediction uncertainties (Chapter 12), and ultimately you should incorporate causality (Chapter 10).

Many problems remain challenging in robustness research:

- Translating a lack of robustness into a robustifying strategy is difficult. Data augmentation can provide a first heuristic solution but there are no guarantees. Also, different shifts require different robustifying strategies.
- Adversarial examples remain relatively mysterious. Can they be avoided and if yes, should they be?
- Synthetic data is often the least costly approach to robustify your model. But, if the data does not resemble the characteristics of real data, training your model on it will be pointless.

Being a machine learning robustness engineer is like being a scientist – it is a constant struggle with nature. Nevertheless, paraphrasing Camus[3], we must imagine machine learning robustness engineers as happy people.

[3]Albert Camus was a French philosopher, author, and journalist who is best known for his existentialist works. In his philosophical essay *The Myth of Sisyphus* (1942), Camus explores the philosophical concept of the absurd – the inherent conflict between the human desire to find meaning in life and the universe's indifference to human concerns. Camus argues that despite the apparent pointlessness of Sisyphus's task, he can still find happiness and meaning in his existence by embracing the absurdity of life and finding fulfillment in the act of rebellion against it. He therefore famously stated that "we must imagine Sisyphus happy". Indeed, connecting these deep thoughts to machine learning robustness is a bit far-fetched...

12. Uncertainty

The Western United States has a problem with water – they don't have enough of it. They even call it *The Dry West*, and if you have ever been to Utah, you know why. Because of this, water management is done for things that require planning, like water storage and agriculture. The Bureau of Reclamation, among others, is concerned with forecasting for each year how much water will be available in the season (from April to July) as measured at certain gauges in rivers and intakes of dams. Since the main water supply in the rivers comes from snowmelt in the mountain ranges and from precipitation, these are the most important features used in forecasting. In the past, hydrologists relied on traditional statistical models, but recently they dipped their toes into machine learning.

Running out of drinking water or losing crops is not an option. That's why decision-makers not only need a forecast, they need to know how certain that forecast is. For example, hydrologists often communicate uncertainty using quantile forecasts (see Figure 12.1): A (correct) 90% quantile forecast means that there is a 90% chance that the actual streamflow will be below the forecast and a 10% chance that it will exceed the forecast. Forecasting the 10, 30, 50 (median), 70, and 90 percent quantiles is common.

In fact, in a world with perfect predictions, all quantiles would "collapse" to the same point, but nature does not allow you to peek into her cards so easily. You have to deal with unforeseen weather changes, measurement errors in mountain ranges, and inadequate spatial coverage. This chapter shows you how to integrate these factors into your machine learning pipeline and qualify your predictions with uncertainty estimates.

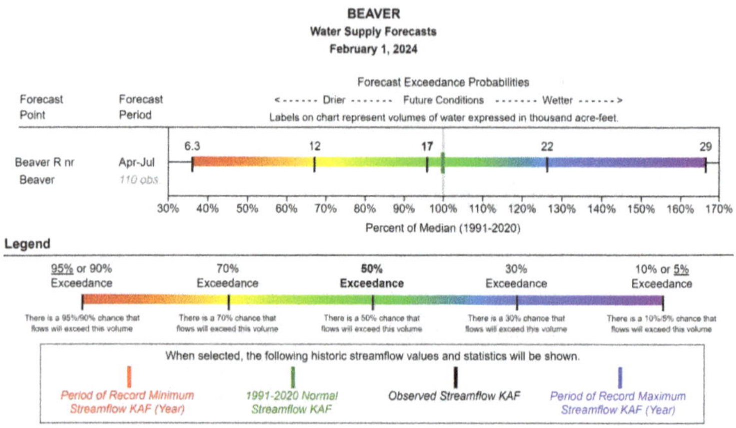

Figure 12.1.: Beaver River water supply forecasts of different "exceedances", which are inverted quantiles (e.g., 90% exceedance = 10% quantile). The February 2024 report was provided by the natural resources conservation service of the U.S. Department of Agriculture, CC-BY (https://creativecommons.org/licenses/by/4.0/).

For a more technical dive into machine learning uncertainty quantification, we recommend [201], [202], [203], and [204], all of which partially inspired our chapter.

> All berries are tasty, but some are dangerous. As a result, berry studies has always been one of the best-funded sciences. So asking Rattle to automate berry classification seemed like a natural step. And indeed, after two weeks, Rattle delivered the world's best berry classifier. The only problem was that nobody trusted the results: The classifier may be right on average, but is it also right for this berry?
>
> 

12.1. Frequentist vs Bayesian interpretation of uncertainty

Say a hydrologist tells you that there is a 10% probability that the April streamflow in the Beaver River will be less than 6.3 thousand acre-feet. How do you interpret that statement? The possible answers describe one of the oldest discussions within statistics: What is a probability?

Frequentist interpretation

Imagine that you examine the water supply (infinitely) many times under similar conditions, then 10% of the time, you would observe water supplies below 6.3 thousand acre-feet. Frequentists view statements about uncertainty as statements about the relative frequency of events in the long run under similar conditions.

The frequentist interpretation of uncertainty has two central conceptual problems: 1) You simply cannot observe an event with an infinite number of repetitions under similar observational conditions, and it is even unclear whether their relative frequencies are well-defined. 2) You are faced with the so-called *reference class problem* – what defines relevant similar conditions?

Bayesian interpretation

The hydrologist believes with 10% confidence that the water supply will be less than 6.3 thousand acre-feet. This belief can be expressed in terms of bets she would be willing to make. For example, the hydrologist would be willing to bet up to 10 cents that the water supply will be below 6.3 thousand acre-feet, receiving 1 Dollar if the water supply is actually below 6.3 and nothing if it's above. To a Bayesian, uncertainty describes subjective degrees of belief. In fact, this subjective belief is constantly updated as new data is observed using the Bayes formula[1].

[1]Bayes formula describes how to update your prior beliefs in light of new evidence. Namely, if $\hat{f}$ is your hypothesis and D is your evidence, you should update your belief with: $\mathbb{P}(\hat{f} \mid D) = \frac{\mathbb{P}(D|\hat{f})\mathbb{P}(\hat{f})}{\mathbb{P}(D)}$.

While the Bayesian approach seems to have solved the limitations of the frequentist approach, it faces its own problems: 1) What should you believe if you have not seen any evidence on the matter? This is the so-called problem of *prior belief.* 2) If uncertainty is a subjective statement about the world, can probabilities be correct? 3) Bayesian probabilities rely on calculating posterior probabilities with the Bayes formula, which is often computationally intractable.

> "Probability is the most important concept in modern science, especially as nobody has the slightest notion of what it means." (Attributed to Bertrand Russell, 1929)

There are also other less established interpretations of uncertainty, such as propensities, likelihoodism, objective Bayesianism, and many others. But whatever understanding of uncertainty you choose, you have to live with its limitations.

Laplace's Demon: Uncertainties in a deterministic world

Does it matter for the frequentist or Bayesian interpretation of uncertainty whether the world is deterministic or not?

Imagine a hypothetical being with infinite computational resources who knows the location and momentum of every particle in the world. Knowing the laws of nature, could it perfectly determine every past and future state of the world? If so, the world would be deterministic, there would be no uncertainty for that being. This thought experiment of a being with infinite knowledge and infinite computational resources is called the *Laplacian Demon.* It would be able to precisely calculate the water supply for April.

Does it matter for the interpretation of uncertainty if such a Laplacian Demon exists? Not really! For frequentists, uncertainties are always defined relative to a reference class of similar conditions specified in a particular language. As long as these similar conditions do not determine the event, the uncertainty remains well-defined. Similarly, for Bayesians, uncertainties arise from *human* computational limitations, incomplete language for the system, insufficient knowledge of the laws, or lack of information about the state of the world. Therefore, both Bayesians and frequentists can reasonably speak of uncertainty at a higher level of description, regardless of whether there is no uncertainty at a lower level for Laplace's hypothetical demon [205]. This chapter deals with uncertainties that arise from the restriction of the prediction task to certain features, inconclusive reference classes, or simply limited human capacities. We lack the expertise to make sophisticated speculations about potential irreducible uncertainties in the quantum world.

12.2. The shared language of uncertainty

While there are great fights about what uncertainty means, there is a relatively broad consensus about the language used to describe uncertainty.

Probability space

The most central concept for describing uncertainty is standard probability theory, which involves:

- A *sample space* Ω that describes all possible outcomes. In our example, an outcome could contain all kinds of information, such as the amount of snow in the U.S., global winds, and the water supply.
- An *event space* $\mathcal{F}$ that describes relevant sets of possible outcomes. One such set might be all outcomes where the water supply is exactly 6.3 thousand acre-feet.
- A *probability measure* $\mathbb{P}$ that assigns a non-negative real value to each element in the event space. For example, the event that the water supply is less than 6.3 thousand acre-feet can be assigned a probability of 0.1 (10%). $\mathbb{P}$ should satisfy the Kolmogorov axioms.

> **i Kolmogorov Axioms**
>
> The Kolmogorov axioms consist of three components: 1) Every event E gets a probability assigned of greater or equal to

zero, i.e. $\mathbb{P}(E) \geq 0$. 2) Ω is in the event space and its probability is 1, i.e. $\mathbb{P}(\Omega) = 1$. This means that one of the outcomes in Ω has to occur for sure. 3) If you unite a countable infinite sequence of events $(E_1, E_2, \dots)$ that are pairwise disjoint, then the union of the events has the same probability as the sum of the individual events, i.e. $\mathbb{P}\left(\bigcup_{i=1}^{\infty} E_i\right) = \sum_{i=1}^{\infty} \mathbb{P}(E_i)$. For example, the event that the water supply is below 6.3 thousand acre-feet (with probability 0.1) and the event that the water supply is strictly above 6.3 thousand acre-feet (with probability 0.9) have no common outcomes, so their joint probability is the sum of the individual probabilities (=1).

Together these three components form a probability space $(\Omega, \mathcal{F}, \mathbb{P})$.

Random variables

Random variables describe quantities based on random events from the event space $\mathcal{F}$. Say you are interested in water supply. Random variables allow you to define a function Y that maps each possible outcome to its corresponding water supply, i.e. $Y : \Omega \to \mathbb{R}$ with $\omega \mapsto$ water supply in ω. This gives you an efficient way to describe events related to water supply, for example, $Y = 6.3$ describes the event where the water supply is 6.3 thousand acre-feet. Throughout this book, we have used random variables to describe the features and the target variable.

Confidence vs credible intervals

Confidence intervals and *credible intervals* quantify uncertainties in a compact way. Say you are not satisfied with a point prediction, and instead you want interval predictions that are likely to cover the true label. Such intervals can indeed be computed. Frequentists will give you confidence intervals, and Bayesians will give you credible intervals. But be careful, they come with different interpretations:

- Frequentist 95% confidence intervals say that with a relative frequency of 95% the true label is contained in an infinite sequence of confidence intervals computed for different data samples. Thus, frequentists see the true label as fixed and the intervals as random variables.

- Bayesian 95% credible intervals say that there is a 95% certainty that the outcome falls into the given interval. Thus, Bayesians see the interval as fixed and the true label as a random variable.

Quantiles

The *q-quantiles* are the cutoff points that divide the probability distribution of a random variable into q equally sized intervals. In practice, the quantiles are relatively easy to estimate:

1. Arrange your data along the variable of interest.
2. Put $\frac{100}{q}$-th of the data with the lowest variable value in the first bin, and so on until all the data is in one of the bins.
3. Compute the average between the highest variable value in the k-th bin and the lowest value in the k+1-th bin. This average is your estimate of the k-th q-quantile.

For example, the first 10% quantile describes the water supply volume where 10% of your observed volumes lie below. This can give you information about the distribution of the data beyond the mean value.

Imprecise probabilities

Standard probability theory assigns each event in the event space $\mathcal{F}$ a value between 0 and 1. This can sometimes be difficult: Think of a hydrologist who is asked for her certainty that the water supply will exceed 12 thousand acre-feet. Instead of giving a point probability of 70% certainty, she might say that her certainty lies between 65-75%. Notice how the information is different from a certainty judgment of 50-90%, even though the average is the same.

Such scenarios are precisely what theories of *imprecise probabilities* are designed for. They allow you to express uncertainty about the probability of a concrete event:

- *Credal sets*, for instance, describe sets of probability distributions rather than a single distribution.
- *Upper and lower probabilities* describe the upper and lower limits of sets of probabilities assigned to a certain event.
- *Interval probabilities* assign each event an interval $[a, b]$ with $a, b \in [0, 1]$ and $a < b$.

There are tons of books on uncertainty quantification, and this small chapter cannot provide the same depth. Instead of providing an exhaustive list of all the concepts and languages used to describe uncertainty, our selection focuses on the most common ones.

12.3. Uncertainty in predictions, performance & properties

Uncertainty matters whenever you estimate things. Scientists particularly care about the uncertainties in the estimations of predictions, performance, and properties.

Predictions

In the hydrology example, we were concerned with estimating the expected water supply. In this case, we care about the uncertainty in *predictions*. What is the probability that the water supply will be less than 6.3 thousand acre-feet? What predicted values cover the true label with 95% certainty? What is the variance in predictions from similarly performing models? Prediction uncertainty is the most common concern of researchers using machine learning. Whenever predictions are the basis for action, investigating prediction uncertainties is a key requirement.

Performance

A hydrologist may wish to compare her model with those of a competitor or with the state-of-the-art. This can be done by evaluating the error on historical/future data that was not used to train the model (i.e., the holdout set). However, this provides only a single estimate of the expected error on other unseen data. Therefore, it is often useful to estimate the test error repeatedly and to define confidence intervals. Whenever scientists want to fairly compare the performance of their model with others, they should be transparent about performance uncertainties.

Properties

In a classic statistical modeling context, hydrologists are interested in the uncertainties of model parameters, assuming that they reflect general properties of interest such as (causal) feature effect sizes. In machine learning, parameters such as weights in neural networks often do not lend themselves to such an intuitive interpretation. There are often neither uniquely optimal nor robust parameter settings. To analyze the properties of interest, such as feature effects and importance, you can generalize the notion of parameters to general properties of interest of the data distribution. These properties and their uncertainties can be efficiently estimated, for example, using targeted learning [206]. Alternatively, the same properties and uncertainties can be estimated using post-hoc interpretability techniques [52]. This allows hydrologists to answer questions about the most effective or important feature of water supply.

12.4. Uncertainty quantifies the expected error

When we talk about uncertainty, we are implicitly talking about errors. How close is the predicted water supply to the true future water supply? How close is your test error to the true generalization error? Does the estimated effect of precipitation on water supply match the true effect?

In general terms, you always have a target quantity of interest T (e.g., true water supply), and your estimate $\hat{T}$ (e.g., predicted water supply) and you want to know how large the error is, measured by some distance function L:

$$\epsilon := L(T, \hat{T})$$

Unfortunately, you don't have access to T. If you did, you wouldn't have to estimate anything at all. Therefore, you usually look at the expected error in your estimation, such as the expectation over datasets used in the estimation.

The reason the expected error can be quantified is the so-called bias-variance decomposition, which works for many known distance functions such as the mean squared error or the 0-1 loss [207]:

$$\underbrace{\mathbb{E}[L(T,\hat{T})]}_{\text{Expected error}} = \underbrace{L(\mathbb{E}[\hat{T}],T)}_{\text{Bias}} + \underbrace{\mathbb{E}[(L(\hat{T},\mathbb{E}[\hat{T}])]}_{\text{Variance}}$$

- **Bias** describes how the expected estimate differs from the true target value T. In a frequentist interpretation, the idea is that if you repeatedly estimate your target you will be correct on average. A biased estimator systematically underestimates or overestimates the target quantity.
- **Variance** describes how the estimates vary, for example, for a different data sample. An estimator has high variance if the estimates are highly sensitive to the dataset it receives.

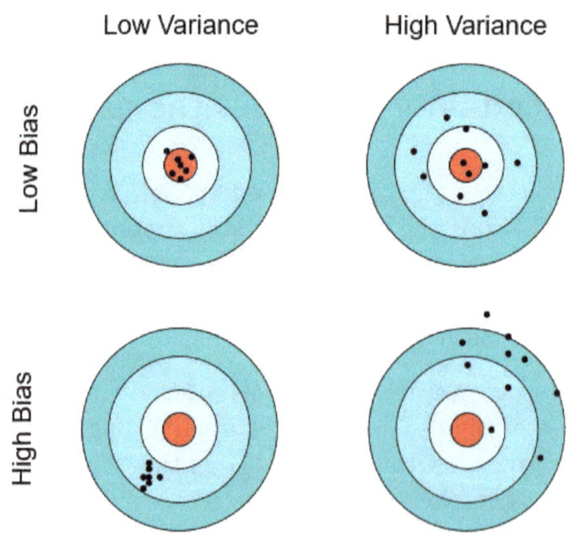

Figure 12.2.: Bias and variance explained intuitively on a dartboard, CC-BY (https://creativecommons.org/licenses/by/4.0/)

But how does bias-variance decomposition help? For many estimators you can prove unbiasedness, meaning the bias term vanishes. Therefore, the expected error can be fully determined from the variance of the estimate. And the variance can estimated empirically or theoretically.

12.5. Sources of uncertainty

Predictions, performance, and properties are the juicy lemonade you get from our whole machine learning pipeline. Whether the lemonade tastes good depends on the water, the lemons, the squeezing technique, and the ice cubes. A lack of quality in any of these has a direct impact on the quality of your lemonade. Similarly, the expected errors in predictions, performance, and properties are due to errors in our task, modeling setup, and data.

Machine learning modeling can be described as a series of steps. Each of these steps can introduce errors that propagate uncertainty in our estimate. The examples and errors shown in the Figure 12.3 focus on *prediction uncertainty*, but performance and property uncertainties can be captured in the same way.

12.5.1. Task uncertainty

Specifying a machine learning task means specifying three things: 1. the prediction target Y; 2. a notion of loss L; 3. the input features $X = (X_1, \dots, X_p)$. In machine learning theory, these three objects are usually considered fixed. In most practical settings, however, they must be chosen carefully.

Prediction target

Often, what you want to predict is vague, making it difficult to operationalize. How do you define the flow of a river? Is one measurement enough? Where to place the sensor? Or should you place multiple sensors and average over many measurements? Vague target variables are especially common in the social sciences. Social

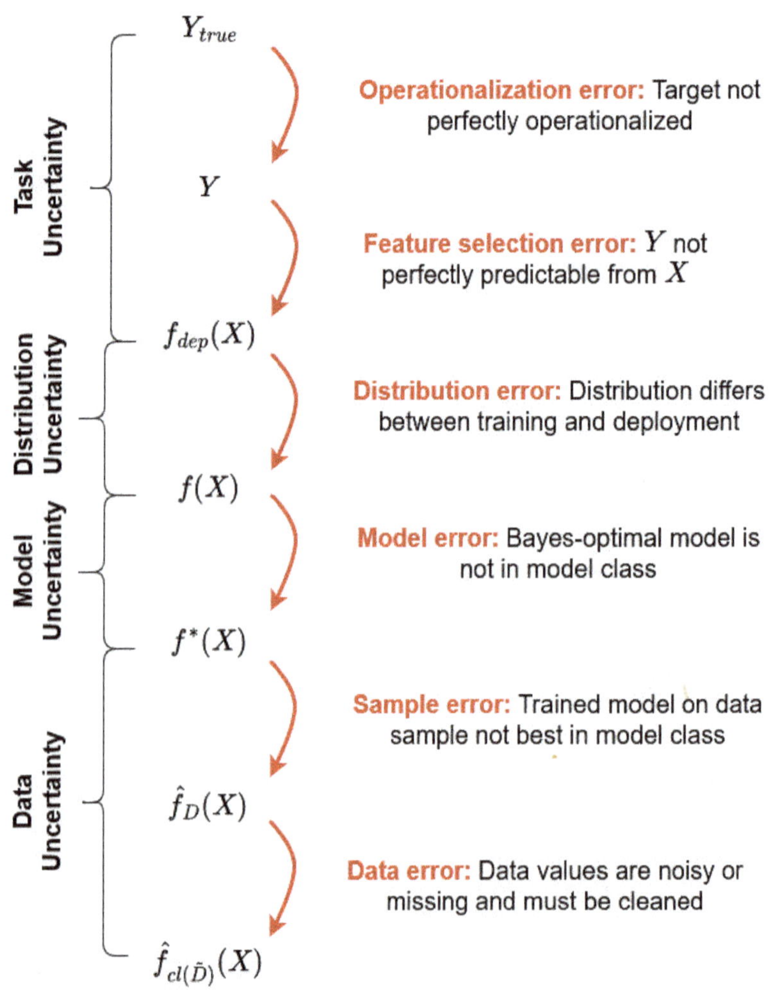

Figure 12.3.: The machine learning pipeline from the perspective of uncertainty. Each step describes an approximation of the true target value, CC-BY (https://creativecommons.org/licenses/by/4.0/)

scientists are often interested in latent variables that cannot be measured directly. Think of concepts like intelligence, happiness, or motivation. For empirical research, these variables need to be operationalized by some measurable proxy. For example, using grades as a proxy for intelligence, money as a proxy for utility, or hours spent as a proxy for motivation. Operationalizations of latent variables are often debatable, to say the least.

The *operationalization error* describes the difference between the true prediction target Y_{true} and its operationalization Y according to some distance function d:

$$\epsilon_{operationalize} := d(Y_{true}, Y)$$

The operationalization error is a big topic in measurement theory [208], but it still needs to be better bridged to scientific practice [209].

Loss

Uncertainty in the target variable leads to uncertainty in the choice of loss function. The loss function is defined over the co-domain of Y and if Y changes, so does the loss L most of the time. An appropriate loss on the money domain $]-\infty, \infty[$ differs from the personal happiness evaluation domain $(1, 2, \ldots, 10)$. Even if there is no uncertainty about Y, there can still be uncertainty about the appropriate notion of loss in a given context. Is it better to define prediction error in water supply in terms of absolute distance or should particularly bad predictions be penalized more using the mean squared error?

Input features

You could use many different features to predict a certain target variable, and each combination of features would result in a different prediction. For example, the snow stations could be placed in different locations. But which is the best prediction? It is complicated! When you have multiple models making predictions based on different features, there is not necessarily one prediction model that is always superior. Each of the models may have its merits in different situations. Constraining the prediction task to a subset of input features X may lead to a feature selection error.

The *feature selection error* describes the difference between the operationalized target Y and the Bayes-optimal prediction $f_{dep}(X)$ according to L in the deployment distribution (see below in Section 12.5.2 our explanation of Bayes-optimal predictors):

$$\epsilon_{feature} := L(Y, f_{dep}(X)).$$

Adding features with predictive value tends to reduce the error. Thus, one might be interested in the difference between the optimal prediction of Y based on feature set X (snow-water equivalent at certain landmarks) and the optimal prediction based on X plus Z (weather forecasts). The *omitted features error* relative to X describes the difference between the optimal prediction based on X and the optimal prediction based on X and Z:

$$\epsilon_{omitted} := L(f_{dep}(X, Z), f_{dep}(X)).$$

12.5.2. Interlude: The Bayes-optimal predictor, epistemic, and aleatoric uncertainty

Given a task (Y, L, X), we can define a central concept when it comes to uncertainty – the Bayes-optimal predictor. It describes a prediction function that takes input features X (e.g., snow-water equivalent, weather forecasts, etc.) and always outputs the best prediction for Y (i.e., water supply). This does not mean that the Bayes-optimal predictor always predicts the correct amount of water supply. It just gives the best possible guess based on X only. As we discussed above, the features we have access to, such as snow-water equivalent and weather forecasts, even if perfectly accurate, will usually not completely determine the water supply. Other factors such as the rock layers below the river, the form of the riverbed, or the lakes connected to the river also affect the water supply. The error made by the Bayes-optimal predictor is the feature selection error $\epsilon_{feature}$ we defined above.

Let's define the Bayes-optimal predictor more formally. It describes the function $f : \mathcal{X} \rightarrow \mathcal{Y}$ that minimizes the generalization error (see Chapter 7 or [210]) and is defined pointwise [2] for $x \in \mathcal{X}$ by:

$$f(x) = \underset{c}{\operatorname{argmin}} \ \mathbb{E}_{Y|X}[L(Y, c) \mid X = x].$$

For many loss functions, you can theoretically derive the Bayes-optimal predictor. For example, if you face a regression task and measure loss with the mean-squared error, the optimal predictor is the conditional expectation of Y given X, i.e. $f = \mathbb{E}_{Y|X}[Y \mid X]$. Alternatively, if you face a classification task and use the $0-1$ loss, the optimal predictor predicts the class with the highest conditional probability, i.e. $\operatorname{argmax}_{y \in \mathcal{Y}} \ \mathbb{P}(y \mid X)$.

Aleatoric uncertainty, epistemic uncertainty, and why we don't like the idea of irreducible uncertainties

In the discussion of uncertainty, a common taxonomy is to distinguish between aleatoric and epistemic uncertainty [204]. Aleatoric uncertainty is said to be irreducible. Even with infinite data on snow-water equivalent, weather, and water supply from the past, and fantastic modeling skills, you might still not be able to forecast water supply perfectly – the remaining error is subject to aleatoric uncertainty. Commonly, aleatoric uncertainty is identified with the feature selection error $\epsilon_{feature}$. Epistemic uncertainty is seen as reducible and stems from your limited access to data, skill in modeling, or computation. For example, if you only have one month of historical data on snow-water equivalent, weather, and water supply, then you have high epistemic uncertainty – collecting more data can reduce this uncertainty.

While we agree that the feature selection error $\epsilon_{feature}$ is of interest, we disagree with calling it irreducible. As mentioned above, including more features allows you to reduce the error in practice.

[2]Technical detail: For $X = x$ with a probability of zero, the term on the right is undefined. Therefore, the Bayes-optimal predictor does not describe a unique function, but an equivalence class of functions that are aligned at x with positive density.

Uncertainty must always be considered reducible or irreducible only relative to the assumptions that are considered fixed, such as the features, the model, or the data. Whether there is a truly irreducible uncertainty is for physicists to decide, not statisticians.

12.5.3. Distribution uncertainty

The next source of uncertainty is the deployment environment. You simply do not know whether the environment in which you will make your predictions will look exactly like the one you have observed. Over the past 100 years, climate change has significantly altered patterns important to water flow: The best water flow predictions in a given setting in 1940 might look different from the best predictions in 2024. It remains uncertain what environmental conditions will prevail on the Beaver River in the coming years: Will Timpanogos Glacier still be there? How much precipitation can be expected in this area? How many extreme weather events such as droughts or floods will occur?

The *distribution error* describes the difference in prediction for a given input x between the Bayes-optimal predictor f with respect to the training distribution (X_{train}, Y_{train}) and the Bayes-optimal predictor $f_{dep}(x)$ with respect to the deployment distribution (X_{dep}, Y_{dep}):

$$\epsilon_{distribution} := L(f_{dep}(x), f(x))$$

We use X, Y, and f instead of $X_{train}, Y_{train}, and f_{train}$ to keep the notation easy to read.

12.5.4. Model uncertainty

The Bayes-optimal predictor is a theoretical construct. The best you can do to get it is to approximate it with a machine learning model. To do this, you need to specify what the relationship between X (e.g. snow-water equivalent) and your target Y (e.g. water supply) might be. The constraints you place on the model will potentially lead to a model misspecification error.

Model misspecification

Choosing the right *model class* for a problem is seen as essential, but what is the right class? What kind of relationship should you expect between snow-water equivalent and water supply? How can snow-water equivalent interact with the weather forecast features? This is difficult! Ideally, you want to choose a model class that contains models that are close to or even contain the Bayes-optimal predictor. Since you usually don't know what the Bayes-optimal predictor looks like, this incentivizes choosing a large model class that allows you to express many possible functions. At the same time, finding the best-fitting model in a large model class is much harder than finding it in a small model class.

Deep neural networks and tree-based ensembles describe highly expressive model classes. They can approximate arbitrary functions well [38], [55], [56] including the Bayes-optimal predictor. Simpler models like linear models or k-nearest neighbors are more constrained in their modeling capabilities.

The bias-variance decomposition provides a good perspective on the uncertainties that can arise from model misspecification. Choosing an expressive model class will reduce the bias in your estimation of the Bayes-optimal predictor. However, it will increase the variance in the estimation process. Conversely, choosing a model class with low expressivity will generally result in a higher bias but lower variance.

Randomness in the hyperparameters selection

One constraint you place on the search for the optimal model within the model class is the *hyperparameters*. The learning rate in stochastic gradient descent, regularizers like dropout, or the number of allowed splits in random forests, all constrain the models that can *effectively* be learned. While the search for suitable hyperparameters can in some cases be automated using AutoML methods

[22], there remains a random element, as trying out all options always remains computationally intractable.

Random seeds and implementation errors

Most complex learning algorithms contain random elements, for example, bootstrapping in random forests and batch gradient descent in deep learning. This randomness is made reproducible with a seed. Running the same non-deterministic algorithm twice with different seeds will produce different predictions. And, there is generally no theoretical justification for choosing one seed over another. Also, implementation errors in machine learning pipelines can affect the learned model.

Model class, hyperparameters, and seeds constrain the set of effectively learnable models

Together, the choices of model class, hyperparameters, and random seeds constrain the effective model class – i.e. the class of models that can be learned under those choices. We call the learnable model closest to the Bayes-optimal predictor the *class-optimal predictor* and denote it by f^*.

The *model error* describes the difference between the Bayes-optimal predictor f within the training distribution (X, Y) and the class-optimal predictor f^*:

$$\epsilon_{model} := L(f(X), f^*(X)).$$

12.5.5. Data uncertainty

Finally, let us look at the ultimate source of uncertainty – the data. You have made measurements in the past and obtained data describing the snow-water equivalent, the weather conditions, and the corresponding water supply. Several things can go wrong:

Sampling

The data you have obtained is only a small sample of the underlying population. With a different sample or more data, your estimate of the label, performance, or property of interest may have looked entirely different.

Uncertainty arising from the data sample can have a large impact on your estimate:

- **Randomness:** You may have been unlucky and taken a sample that contains many outliers. For example, you took measurements on random days, but they happened to be very rainy days.
- **Data size:** Your data sample may be too small to be representative of the underlying probability distribution. For example, one measurement per month is not enough for your model to learn the relevant dependencies.
- **Selection bias:** Your sample may be biased, resulting in a non-representative sample. For example, you may have taken measurements only on dry days to avoid getting wet.

The *sample error* describes the difference between the prediction of the class-optimal model f^* and the learned model $\hat{f}_D$ based on the complete dataset D:

$$\epsilon_{sample} := L(f^*(X), \hat{f}_D(X))$$

In the literature, this error is sometimes referred to as the *approximation error* [204], [211], but we found the term *sample error* more informative about the nature of this error.

Measurement errors

All data are derived from measurements, or at least it is useful to think of it that way. Your water flow sensors make a measurement. Tracking the amount of snow is a measurement. Even just taking a picture of that snow is a measurement. A measurement error is the deviation of the result of a measurement from the (unknown) true property. It is inevitable to make measurement errors.

- **Measurement error in the target** introduces a bias. Think of human transcription errors in water flow data.[3] There are two directions for this bias: it could be above or below the true property of interest. You might get lucky and biases cancel out on average but that is unlikely, especially if the error is systematic.
- **Measurement errors in the features** may or may not introduce a bias in the predictions. If there is a random error in the measurement of the snow-water equivalent, it can be washed out during training. Systematic measurement errors in the input features can have more serious consequences: If the snow-water equivalent measurement contains large errors, the model may rely on a proxy feature such as streamflow in the previous month.

Missing data

In an ideal dataset, no cell is empty. Scientific reality is often less tidy. Some values in your rows will look odd: The water flow on March 22 was *infinite*? The current snow-water equivalent is *not a number (NaN)*? The date human-entered record is *November 32 in 2102*? What should we do if some values are missing or undefined?[4]

There are several reasons why data may be missing:

- **Missing completely at random (MCAR):** The missing value mechanism is independent of X and Y. For example, the company that tracks the snow-water equivalent has a server problem and therefore the value is missing.
- **Missing at random (MAR):** The probability of missingness depends on the observed data. For example, if the guy who sends you the snow-water equivalent sometimes misses it on Thursdays because it is his date night.

[3] Many of these measurements from gauges and snow stations are fully automated. But for the sake of illustration, let's assume they are recorded by humans.

[4] We do not talk about whole entries missing here, because this typically leads to problems of representativeness and is covered in Chapter 7.

- **Missing not at random (MNAR):** The probability of missingness depends on the missing information. For example, if you don't get a snow-water equivalent measurement during a snowstorm because the staff can't reach the sensors.

Measurement errors and missing values lead to another estimation error

Measurement errors and missing values make the data set you deal with in practice look different from the complete data set with unbiased values D you consider in theory. The *data error* describes the difference between the learned model $\hat{f}_D$ based on the complete and accurate dataset D and the learned model $\hat{f}_{cl(\tilde{D})}$ on the dataset $\tilde{D}$ that contains measurement errors and missing values that need to be cleaned up using operations cl:

$$\epsilon_{data} := L(\hat{f}_D(X), \hat{f}_{cl(\tilde{D})}(X)).$$

12.6. Quantifying uncertainties

Now we have fancy names for all kinds of sources of uncertainty and the errors they represent. But how can we quantify these uncertainties?

> i Uncertainties are hard to disentangle in practice
>
> Above we tried to disentangle all kinds of uncertainties: What is the difference between the true target and its operationalization? What is the difference between the Bayes-optimal predictor and the class-optimal model? What is the difference between the class-optimal model and the model learned on an imperfect dataset? In practice, however, you do not have access to the true target, the Bayes-optimal predictor, or the class-optimal model. You always have to run the whole process. The different errors are theoretically well-defined and useful for thinking about minimizing uncertainties, but

in practice, the uncertainties are mixed together.

12.6.1. Frequentist vs Bayesian uncertainty quantification

Frequentists and Bayesians approach uncertainty quantification differently.

Frequentists have a simple default recipe – they like repetition. Say you want to quantify the uncertainty in your estimate:

1. Assume, or better *prove* that your estimation is unbiased, i.e. in expectation it will measure the right target quantity.
2. Run the estimation process of the target quantity (i.e., predictions, performance, or properties) multiple times under different plausible conditions (e.g., task conceptualization, modeling setup, or data).
3. Because of the bias-variance decomposition and the unbiasedness from Step 1, the uncertainty comes exclusively from the variance of the estimates.

This approach has limitations. For example, you cannot prove that your task conceptualization is unbiased. And it can be difficult to come up with multiple plausible conditions for estimation. For frequentist uncertainty quantification, you need confidence in your domain knowledge. But then it is a feasible approach that can be applied post-hoc to all kinds of tasks, models, and data settings. Note, however, that some methods, such as confidence intervals, require additional assumptions that the errors are IID, homoscedastic (i.e., remain the same across different data instances), or Gaussian.

Like frequentists, Bayesians have a standard recipe – they just love their posteriors. Say you want to quantify the uncertainty in your estimate:

1. Model the source of uncertainty (e.g. distribution, model, or data) explicitly with a random variable and a prior distribution over that variable.

2. Examine how the uncertainty propagates from the source to the target quantity (i.e., predictions, performance, or properties).
3. Update the source of uncertainty, and consequently the uncertainty in the target, based on new evidence (e.g., data) using Bayes' formula.

In terms of uncertainty quantification strategies, we focus mainly on frequentist approaches. Bayesian uncertainty often needs to be baked in from the start, including a refinement of the optimization problem, while frequentist uncertainty is often an easy add-on. But honestly, another reason is that we are just more familiar with frequentist approaches. For a detailed overview of Bayesian uncertainty quantification strategies in machine learning, check out [212] or [213].

12.6.2. Directly optimize for uncertainty

The simplest approach to uncertainty quantification is to predict uncertainties rather than labels. This is often done in classification tasks. Instead of simply predicting the model class with the highest output value, the output values are transformed with a softmax function. The softmax function transforms arbitrary positive and negative values into values that look like probabilities; they are non-negative and sum up to one. When predicting probabilities, you also need different loss functions – the default choices here are cross-entropy and Kullbach-Leibler divergence.

For regression tasks, optimizing directly for uncertainty is less straightforward. Typically, the probability of any individual target value is zero. However, you can estimate the density function directly. This is usually done with Bayesian approaches such as Bayesian regression, Gaussian processes, and variational inference [212]. Here, you explicitly model the uncertainty about the optimal model, either via a Gaussian distribution over functions or parameters, and update the model in light of new data.

More technically, this means that you define a prior distribution $p(\hat{f})$ over the set of models $\hat{f} \in F$. Based on this, you can compute the posterior distribution of the models given your data $p(\hat{f} \mid D) = p(D \mid \hat{f})p(\hat{f})/p(D)$. Finally, this allows us to estimate the target in a Bayesian manner by

$$p(y \mid x, D) = \int_{\hat{f}} p(y \mid \hat{f})p(\hat{f} \mid x, D)\, d\hat{f}.$$

This integral has to be solved numerically again [212].

While you get numbers from direct estimation that look like probabilities, they are often difficult to interpret as such. Especially because they are not calibrated – i.e., they do not match the probabilities in the true outcomes. We discuss how to calibrate probabilities in Section 12.8.

12.6.3. Task uncertainty is hard to quantify

As a reminder, operationalization error describes the difference between the true target (e.g., intelligence) and its operationalization (e.g., IQ test). Quantifying this difference is difficult. There are several challenges: First, the true target is sometimes not directly measurable. Second, even if you had access to the true target and the operationalization, they wouldn't be on the same scale, making it hard to define a distance function.

What can be done, however, is to look at the coherence or correlations between different operationalizations [202]. For example, in intelligence research, the correlation between different operationalizations of intelligence led to the so-called *G-factor*, which became the real target that researchers wanted to measure with intelligence tests [214]. Operationalizing the G-factor through multiple specific operationalizations and tests allowed researchers to quantify the operationalization error of specific intelligence tests.

Remember that the feature selection error describes the difference between the operationalized target Y (like water supply) and the Bayes-optimal predictor given a feature set X (e.g., snow-water equivalent). It is more tangible to quantify than the operationalization error but it is still difficult. The best proxy for the expected overall feature selection error is the test performance of a well-trained model. For example, if you squeeze the highly engineered water supply prediction model down to a certain test error, the remaining error may arguably be the feature selection error. If you have squeezed out all the predictive value contained in the available features, techniques like cross-validation allow you to additionally estimate the variance in this error.

Quantifying the expected feature selection error for individual instances x (e.g., the snow-water equivalent and weather forecasts for a given day) is even more difficult. Even if the prediction model as a whole is close to the Bayes-optimal predictor, it may be far off for individual instances x. To see how it might still be possible to quantify the error, check out our discussion of Rashomon sets below.

The omitted feature error can be quantified using interpretability techniques such as *leave-one-covariate-out (LOCO)* [114] or *conditional feature importance* [215]. These techniques also allow to quantify uncertainty but always require labeled data of (X, Z).

12.6.4. No distribution uncertainty without deployment data

To estimate distribution uncertainty, we need to know how the training distribution differs from the deployment distribution, or at least have some data from both distributions. If we have access to data, we can compute the respective errors directly. Say one researcher only has access to Beaver River data from 1970 to 1980 while the other has data from July 2014 to July 2024. Both use machine learning algorithms to get the best possible models. Then, they can compare their predictions, performance, or properties on the most recent data to assess the expected distribution error.

Often, we are interested in deployment uncertainty but do not have deployment distribution data. Physical models can the be used to simulate data from different potential distribution shifts. A more detailed discussion on data simulation and augmentation can be found in Chapter 11.

12.6.5. Rashomon sets – many models are better than one

Imagine meeting with ten water supply forecasting experts. Each knows the historical data and has insights into snow-water equivalent and weather forecasts. But they come from different modeling schools, there are physicists, statisticians, machine learners, and so on. Each of them gives you their honest estimate of the water supply in Beaver River for June – but their predictions differ. Not much, but enough to matter. What should you do? There is no good reason to think that any of the ten experts will stand out. The intuitive strategy would be to predict the average (or the median if there are outliers) and analyze the variance in the experts' opinions to assess the uncertainty in your prediction.

Why not just apply this reasoning to machine learning? Each of the ten experts can be thought of as a different learning algorithm, turning past data into models and current data into predictions. Say you have trained ten models from different model classes, using different hyperparameters and random seeds, all of which perform similarly well. Then this set of models is called a *Rashomon set.*

Say there is a well-trained model $\hat{f}_{ref}$ as a reference point and you fix a certain admitted error δ. Then, $S := \{\hat{f}_1, \dots, \hat{f}_k\}$ is a Rashomon set to $\hat{f}_{ref}$ if for all $i \in \{1, \dots, k\}$ holds that the performance of $\hat{f}_i$ is no worse than δ, i.e.

$$\mathbb{E}[L(\hat{f}_i(X), Y)] - \mathbb{E}[L(\hat{f}_{ref}(X), Y)] \leq \delta$$

.

If you assume that your collection of models in the Rashomon set comes from an unbiased estimation process of the optimal model, you can estimate the expected model error by the variance of the predictions in the Rashomon set, i.e.

$$\hat{\mathbb{V}}_{f^*}[f^*(x)] = \frac{1}{k-1} \sum_{\hat{f}_i \in S} L(\bar{\hat{f}}(x), \hat{f}_i(x)).$$

Assuming a certain distribution of errors (e.g., t-distribution) and IID errors in the Rashomon set, you can use the variance to define confidence intervals. For example, the α confidence interval for the predictions would be:

$$CI_{\hat{Y}} = [\hat{f}_{ref}(x) \pm t_{1-\alpha/2}\sqrt{\hat{\mathbb{V}}_{f^*}[f^*(x)]}]$$

As with all strategies presented here, Rashomon sets can be used to quantify not only the uncertainty in predictions but also the uncertainty in estimated performance and properties. For example, Rashomon sets have been used to quantify the uncertainty in estimating feature importance [105] and feature effects [110]. Similarly, even if the predictions are already probabilities, as is common in classification models, Rashomon sets can be used to quantify higher-order uncertainties. For example, you can obtain imprecise probabilities in the form of intervals by taking the highest and lowest predicted probabilities in the Rashomon set as interval bounds.

Interestingly, in a very Rashomon-set fashion, ensemble methods like random forests implicitly provide an uncertainty quantification of the model error [216]. Each decision tree in the forest can be seen as one prediction model. Similarly, dropout in neural networks can be described as an ensemble method with implicit uncertainty quantification [217].

12.6.6. The key to data uncertainty: sampling, resampling, repeated imputation

The first error we mentioned in the context of data uncertainty is the sample error. In an ideal world, it would be easy to quantify: 1. Take many independent, unbiased samples of different sizes from the same distribution. 2. Quantify the variance in the resulting estimates (i.e., predictions, performance, or properties). This is nice in theory or in simulation studies. But in real life, data is usually both valuable and scarce. You want to use all available data to make your estimates as accurate as possible, not just waste 90% on uncertainty quantification.

Resampling can quantify sample uncertainty, but it can systematically underestimate it

Statisticians have therefore developed smart strategies to avoid wasting data under the umbrella term *resampling*. The idea: make your best estimate based on the entire dataset. To quantify the sampling uncertainty of that estimate, you study the variance when you repeatedly re-estimate the quantity of interest on different subsets of the full dataset.

There are several approaches to resampling:

- *Bootstrapping* involves repeatedly drawing samples with replacement.
- In *subsampling*, you repeatedly draw samples without replacement.
- With *cross-validation*, you split your data into k-junks of equal size. *Cross-validation* is particularly useful when you need separate data splits for training and estimation, as in performance estimation (see Chapter 7).

But be careful, with resampling approaches you run the danger of underestimating the variance and consequently the true uncertainty. The reason is that the samples you draw are not really independent. They come from the same overall dataset and share many instances. There are several strategies to deal with the underestimation, such as variance correction strategies like the *bias-corrected and accelerated bootstrap* [218]. For performance estimation, Nadeau and Bengio [219] suggest several correction factors to tackle underestimation.

Strategies for noisy or missing data

Measurement errors and missing data may look different from the outside, but when you think about it, they are similar. A very noisy feature value can just as well be seen as missing. And, a missing value that is imputed can also be considered noisy. This similarity is reflected in the shared data-cleaning strategies for dealing with them:

- **Remove the feature:** If a feature is generally super noisy and often missing, we recommend removing it. This way, you can ignore the data error but potentially have a higher feature selection error. Therefore, removing features may be a bad idea if the feature is highly informative about the target. For example, removing the snow-water equivalent from your forecast model will decrease forecast performance.
- **Remove individual data points:** If the values of a feature are missing *completely at random* in only a few cases, it may be fine to remove the affected data points. This way, no data uncertainty is introduced. Note that removing data reduces your data size and therefore increases the sampling error. Also, if the missingness or noise is not random, you may introduce selection bias into your dataset.
- **Impute missing values:** If the missingness is random (MAR), we recommend using data imputation strategies. Ideally, the imputation process takes into account your domain knowledge and data dependencies. Note that imputing missing values based on domain knowledge is itself only a guess. The imputation will sometimes be incorrect.
- **Explicitly model uncertainty:** If the values of a feature are moderately noisy and you have a guess about how the noise is distributed, we recommend that you model the noise explicitly. This way, the uncertainty can be quantified by propagating the noise forward to the target estimate.
- **Improve the measurement:** If the missingness is not at random (MNAR), the only good strategy is to get rid of the reason for the missingness and collect new data.

12.7. Minimizing uncertainty

Great, so now you have identified uncertainties and found ways to quantify them – but how do you minimize them? In most contexts, you want to reduce uncertainty: For example, when the water supply model tells you that it is uncertain whether there will be a drought or a flood in the Beaver River. How should you take precautions if the uncertainty is high?

Most, if not all, uncertainties can, in principle, be minimized. Some are just harder to minimize than others. While the goal is to minimize that total uncertainty, the only strategy for doing so is to minimize the individual uncertainties that arise from different sources:

- **Task uncertainty**
  - To minimize the expected operationalization error, put considerable effort into an appropriate operationalization. For example, placing only a single sensor in the river might be a bad idea if the water flows vary along the river.
  - To minimize the expected feature selection error, select features that are highly predictive and contain little noise. Omit features that do not add new information to your model.
- **Distribution uncertainty**
  - To minimize the expected distribution error, collect or simulate training data that is representative of the deployment distribution. For example, training your model only on data from the 1970s may be a bad idea given the changed distribution of extreme weather events.
- **Model uncertainty**
  - To minimize model error, choose an appropriate inductive bias in modeling. Strongly constraining the model class and testing a few hyperparameters will result in a high bias. Choosing too broad a model class and unconstrained hyperparameter search will result in high variance. To minimize uncertainty, use your domain knowledge to constrain your model search to a minimal but

sufficiently expressive model class to capture the dependency in the data and plausible hyperparameter settings [220]. In water supply forecasting, it is recommended to include physical constraints because they constrain the appropriate model class and consequently reduce model uncertainty [221].
  - To find an appropriate model class and hyperparameters, two strategies can be helpful: Either start with simple models and increase complexity until there is no performance gain or start with a complex model and sequentially decrease complexity until there is a performance drop.

- **Data uncertainty**
  - To minimize the expected sample error, the best strategy is to collect more data. Active learning can help you select the data that will reduce uncertainty the most (see the active learning in Chapter 11). However, active learning is impossible in domains like water supply forecasting, where control over the data-generating mechanism is limited.
  - Collect multiple labels for an instance to minimize the expected data error in the face of label noise. If there is feature noise, either reduce the noise through improved measurement or replace the feature with an equally predictive but less noisy feature. For example, move a snow sensor from a mountain area with a lot of skiers to a quieter area.
  - If the problem is missing data, investigate the cause of the missingness: in many cases, the missingness can be counteracted by a careful measurement, like replacing an unreliable sensor. If the missingness cannot be counteracted, consider removing the feature.

12.8. Calibrating uncertainty measures

Say you have studied the sources of uncertainty, quantified them, and perhaps even minimized them. Of course, you want to interpret the values in your uncertainty estimates as true probabilities. But there is often a problem with these "probabilities". The fact that your uncertainty estimate has a 90% confidence does not necessarily mean that the actual probability is 90%. Probabilities can be miscalibrated – the estimated probabilities don't match the true outcome probabilities.

What does calibration mean in more formal terms? Say you are in a regression setting, such as the water supply forecasting problem. We denote the prediction by $\hat{Y}$ and the uncertainty estimate for $\hat{Y}$ with confidence c by $u_c(\hat{Y})$. We say that the uncertainties are perfectly calibrated if for all $c \in [0, 1]$ holds:

$$\mathbb{P}(Y \in u_c(\hat{Y})) = c.$$

For example, the hydrologist's certainty of 10% that the water supply in the Beaver River will be less than 6.3 thousand acre-feet is calibrated if the actual water supply in the Beaver River is less than 6.3 thousand acre-feet only 10% of the time. Calibration can be similarly defined for classification tasks.

Calibration plots allow you to assess whether probabilities of a classifier are calibrated or not. You compare the predicted and quantified uncertainty with the empirical frequency. For this you pick an outcome class and bin the data points by output score, for example into 0-10%, 10-20%, ..., 90-100%. On the x-axes, you plot the confidence bins and on the y-axes the empirical frequencies with which the class matches the chosen class. Figure 12.4 shows an example of a calibration plot, where we compare the mean probabilities with the actual probabilities. If the plot shows a boring linear curve with a slope of 1, congratulations, your model is perfectly calibrated. Evaluating calibration using relative frequencies shows the close relationship between calibration and a frequentist interpretation of probability.

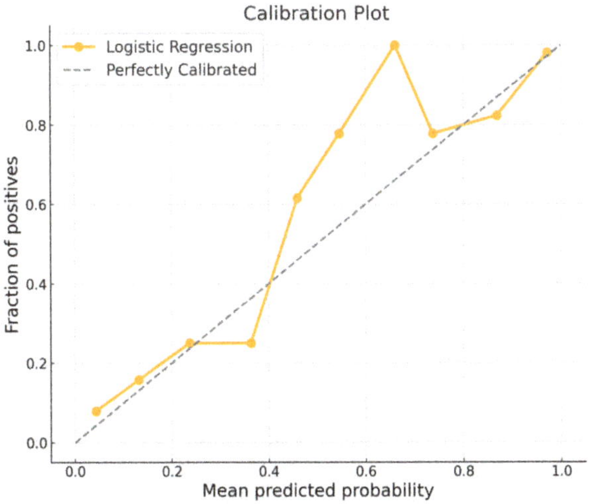

Figure 12.4.: Example of a calibration plot: The logistic regression model is not perfectly calibrated, especially for predicted probabilities between 0.5 and 0.7. But it could be worse.

Miscalibration is not just an occasional problem. It is a fair default assumption that any uncertainty estimate is miscalibrated:

- **Too short prediction intervals:** For the water flow prediction example, we were interested in prediction intervals that cover 80% of the output. The interval was created by predicting the 10% and 90% quantiles. The interval width expresses uncertainty: The larger the interval, the more uncertain the water flow. However, quantile regression has the problem that quantiles are often drawn toward the median [222], an effect that becomes stronger the closer the desired quantile levels are to 0% and 100%. As a result, intervals based on quantile regression tend to be too short and underestimate uncertainty.
- **Bootstrapping also undercovers:** Remember that in bootstrapping you repeatedly sample data with replacement from the training data. Because the bootstrapped datasets are highly correlated, they underestimate the true

uncertainty unless you correct for this bias.

- **Bayesian models rely on assumptions:** In theory, Bayesian models propagate uncertainty perfectly. You could look at the predictive posterior distribution and, for example, output credible intervals. But they are calibrated (in the frequentist sense) only if your assumptions about likelihood, priors, and so on are correct. Hint: they are unlikely to be correct.

For each miscalibrated uncertainty quantification method, there are several suggestions on how to fix it. For example, to fix miscalibrated probability outputs, you could use post-processing methods such as Platt's logistic model [223] and Isotonic Calibration. A more general framework for dealing with all these problems is conformal prediction, which can make any uncertainty measure "conformal" – with coverage guarantees.

Conformal prediction, a set-based approach to uncertainty

Standard uncertainty quantification treats models and predictions as fixed and returns the uncertainties. Conformal prediction takes the opposite approach: The modeler must specify the desired confidence level (related to uncertainty), and a conformal prediction procedure changes the model output to conform to the confidence level. Consider water supply forecasting: Normally, you would simply predict a scalar value and provide the quantified uncertainty estimate in some form. With conformal prediction, you first set a confidence level and then modify your confidence interval to cover the true value with the desired confidence. Thus, conformal prediction comes with a *coverage guarantee* – at least for data that is exchangeable, the prediction set corresponds to the desired confidence level. But beware, the coverage guarantee is often marginal, i.e. it only applies to the average of the data.

Conformal prediction is more than a simple algorithm, it is a whole framework that allows to conformalize different uncertainty scores:

- Turn a quantile interval into a conformalized quantile interval [224]).
- Turn a class probability vector into a set of classes [225].
- Turn a probability score into a probability range (Venn-ABERS predictor) [226].

There is a simple recipe for conformal prediction – *training*, *calibration*, and *prediction*.

1. **Training:**
    - You split the training data into training and calibration data.
    - Train the model using the training data.
2. **Calibration:**
    - Compute uncertainty scores (also called nonconformity scores) for the calibration data.
    - Sort the scores from certain to uncertain (low to high).
    - Decide on a confidence level.
    - Find the quantile where $1 - \alpha$ of the nonconformity scores are smaller.
3. **Prediction:**
    - Compute nonconformity scores for the new data.
    - Select all predictions that produce a score below $\hat{q}$.
    - These predictions form the prediction set/interval.

Nothing comes for free and neither does conformal prediction. The "payment" is the additional calibration data you need, which can complicate the training process. Small calibration sets will lead to large prediction sets, possibly to the point where they are no longer useful. Furthermore, the calibration data must be interchangeable with the training data, which can be a problem with time series data, for example. However, there are also conformal procedures for time series that rely on a few more assumptions. Otherwise, conformal prediction is a very versatile approach.

Introduction to Conformal Prediction

To learn more about conformal prediction, check out the book Introduction to Conformal Prediction with Python[a].

[a] https://christophmolnar.com/books/conformal-prediction/

12.9. Understand uncertainty to gain knowledge

In this chapter, you learned about the many different sources of error in machine learning pipelines and how they introduce uncertainty into your estimates. While uncertainty can be disentangled in theory, it is difficult to disentangle in practice – you can usually only quantify the overall uncertainty. Nevertheless, it is helpful to separate uncertainties conceptually, especially if you are looking for strategies to minimize uncertainty.

Uncertainty should not be seen as just an add-on. Uncertainty quantification is essential, especially if you want to gain new knowledge with machine learning or act on machine learning predictions. To get a realistic picture of the true uncertainty, all sources of uncertainty should be considered

When it comes to quantifying uncertainty, we recommend that scientists be pragmatists. You can use Bayesian techniques if you have domain knowledge that allows you to set a reasonable prior. Otherwise, frequentist approaches to uncertainty quantification can provide you with (calibrated) uncertainties under modest assumptions.

Uncertainty is closely related to many chapters in this book:

- Theoretically, the roots of uncertainty quantification lie in our theory of generalization from Chapter 7. Concepts such as generalization error, cross-validation, or biased sampling can be found in both chapters.
- Many approaches to minimizing uncertainty rely on domain knowledge (see Chapter 8). Reliable domain knowledge can significantly minimize task, distribution, model, and data uncertainty.
- The uncertainty framework presented here is quite general. It can be applied to interpretability techniques (see Chapter 9) and even to causal effect estimation (see Chapter 10).
- Uncertainty has close ties to the robustness in Chapter 11, especially when it comes to distribution uncertainty and strategies to reduce data uncertainty like active learning. Conceptually, the link is that robustness gives guarantees for worst-case error, while uncertainty gives guarantees in expectation [28].

We hope we have minimized your uncertainty about uncertainty with this chapter...

13. Reproducibility

The beauty of code: Once written, you can use it as often as you like. If set up correctly, the same code applied to the same data will produce the same results. Ideally, training your machine learning model works the same: at the click of a button, you can run your code again and get the exact same machine learning model. This is known as computational reproducibility, and it comes with many advantages[1].

- Reproducibility makes it easier to track changes in your project.
- You can reproduce your work. Comes in handy when reviewer 2 asks you to retrain your model.
- People will enjoy working with you and your beautiful code base.
- Other researchers can build on your work.
- Reproducibility allows you to put a model into production later.

But making your code reproducible is much harder than it seems. Computational reproducibility is inherently unstable due to ever-changing computing environments and the unique challenges of machine learning.

> **i** Reproducibility versus replicability
>
> You may have heard about the replicability crisis in the social sciences [227]: Many research findings couldn't be replicated. Reproducibility focuses on a narrower goal: getting the

[1] https://book.the-turing-way.org/reproducible-research/overview/overview-benefit

> same results (e.g. the same model) using the same code on the same data. For an overview of definitions, see the Turing Way Book[a] [228].
>
> [a]https://book.the-turing-way.org/reproducible-research/overview/overview-definitions

Reproducibility is not binary, but it is a spectrum. This chapter will help you move up that spectrum by highlighting the unique challenges that machine learning poses to reproducibility. Let's dive in.

> The tornado prediction model worked like a charm for many years. Until it didn't. One hot summer day, the server had a meltdown and the model was gone. Rattle went into early retirement with a wealth of money, but her original learning algorithm was still sitting on an old laptop. Well, should be a simple retraining task. Or so they thought. They were confronted with a strange folder structure, mysterious file names, and unclear instructions. Other Ravens' code can be such a pain.

13.1. Making any coding project reproducible

Reproducibility is a challenge for any coding project even without machine learning. This chapter focuses more on the machine learning-specific challenges, but it wouldn't be complete if we didn't mention some general tips and tricks for reproducibility [229]:

- Create a clear folder structure.
- Use good names for files, folders, and functions.
- Document your project, including a README, metadata, and code documentation.
- Use version control software such as git to track changes to code and text.
- Stabilize the computing environment and software using environment management tools such as Conda and virtualization tools such as Docker.
- Automate computations, e.g. Makefile, workflow stuff, and have all steps reproducible with code.
- Publish your work.

The more of these tips you follow, the more reproducible and satisfying your project will be. However, it may require learning new tools like git and adopting new habits like documenting your code – this is an initial investment of time and effort, but it will pay off in the long run. Your future self will thank you.

Now let's move on to the machine learning-specific challenges.

13.2. Handling non-deterministic code

Running the same code twice can produce different results due to randomness. Machine learning involves a lot of randomness: random weight initialization, stochastic gradient descent, random forests, and random data splitting. It is not a bug, it is a feature: Randomness is an inherent driver of learning. For example, randomly splitting data into training and testing simulates "drawing from the same distribution" which is a crucial element of generalization (see Chapter 7) and stochastic gradient descent implicitly regularizes the model through its random nature [230].

To introduce randomness into the deterministic world of programming logic, machine learning software relies on "random number generators". These generators are not truly random – they are based on pseudorandom processes that can be controlled by setting an initial random seed. A random seed initializes the random number generator, and all subsequent "random" numbers are deterministic.

This random seed is also your key to reproducibility for code with randomness. Without a random seed, training a machine learning model will produce different models every time. At least it is unlikely to get the same results twice. But by setting a random seed, you make the random number generation reproducible. There are some exceptions: When running multiple processes in parallel, setting a random seed may still not guarantee deterministic results if the order of execution is inconsistent.

Besides your computer's random number generator, there are other reasons why running the same code twice may produce different results, even if you have followed all the tips, like making sure you are using the same software:

- Operations on GPUs can be non-deterministic, even simple ones like a sum (see this StackOverflow question[2]).
- External systems can change over time (e.g. API calls).
- Floating point arithmetic can vary between different hardware and software platforms.

13.3. Jupyter Notebooks

Jupyter Notebooks are a blessing and a course to research.

> **i Jupyter Notebook**
>
> The term refers to both the software and the document, just as "Excel" can refer to both the program and a single spreadsheet. Jupyter Notebooks – the software – is essentially an HTML application where you can manage multiple notebooks (the documents), create new ones, delete, edit, and run them. A notebook – the document – is a collection of "cells". A cell

[2] https://stackoverflow.com/questions/50744565/how-to-handle-non-determinism-when-training-on-a-gpu

can contain markdown, code, or plain text. Markdown cells are rendered so that you can structure your notebook like a document, with titles, bold text, and other formatting. You can execute code cells, and the results are embedded in the document, whether it is a plot, code warnings, or a printout.

Notebooks let you experiment, quickly prototype ideas, and explore data; they encourage documentation; they are great for reporting results. But they make reproducibility difficult. In his provocative talk "I Don't Like Notebooks"[3], Joel Grus specifically criticized "hidden states" that make reproducibility difficult:

- You can run the cells in any order.
- You can delete cells, but the variables that were created remain in the workspace.
- You can change a cell's content while only running the old version.

Imagine you're working on a machine learning project in a Jupyter Notebook. You decide to standardize features before training but forget to run the updated cell. You save the model, but now the code doesn't match the model, and re-running the notebook produces a different model.

In another scenario, you accidentally delete the cell that generates a weight vector shortly after writing the code. But because you executed the cell, the vector is in memory. You finish the project, check your latest changes into version control, and call it a day. The next person to work on the project gets an error about the missing weight vector.

There are even more problems with notebooks: While you can put them under version control, they are quite verbose because they store all HTML output, and actually comparing code changes is no fun like this.

[3] https://docs.google.com/presentation/d/1n2RlMdmv1p25Xy5thJUhkKGvjtV-dkAIsUXP-AL4ffI

It is possible to make a research project reproducible, even if it relies on notebooks. But you have to be very strict about running them *linearly* from the first cell to the last, and always with a fresh Python session.

13.4. APIs and proprietary software

Training a machine learning algorithm on someone else's hardware can make your life a lot easier: You don't have to worry about hardware and software installations. Hundreds of machine-learning-as-a-service platforms allow you to upload your data and train a model. The problem: lack of reproducibility. The company behind the machine learning software can change its software without informing you. And even if they don't: You may not have enough control over the workflow to make it fully reproducible for others. It may not be transparent what algorithms, settings, and software versions the platform uses. Or the company may simply go out of business and you lose access to your training setup.

This problem has gotten much worse with generative AI, especially with large language models like ChatGPT. Whether you are studying ChatGPT for bias or using it to label data: The company behind it is constantly updating the models, and once a model is retired, no one can reproduce your research.

13.5. Hardware-specific challenges

Even if you control the server or use your laptop and have followed all the reproducibility recommendations, your project may still not be at 100% reproducible. At least when it comes to running your code on different hardware. The problem is that we can't completely abstract the training and so on from the hardware. Increasingly, machine learning, especially deep learning, relies on more specific hardware, such as NVIDIA GPUs and Google TPUs. The more hardware-specific your setup, the harder it is for others to reproduce your models.

13.6. Inaccessible data

Data can be inaccessible for a number of reasons: it may be too large to share, there may be privacy concerns (such as patient data), or it may be proprietary. This makes reproducibility difficult. But even making it reproducible within the project itself can be difficult, especially with large datasets. A possible "incremental" solution would be to share a subset of the data or simulated data.

> **i** Reproducible versus reusable
>
> Reproducibility and reusability are not the same thing: A research result can be reproducible without being reusable. You may be able to reproduce someone else's project but if, for example, many things are hard-coded, it will take a lot of effort to adapt it to your use case.

13.7. Reproducibility and other issues

Reproducibility is about more than just making sure that others, including your future self, can reproduce your work. It relates to many other aspects we cover in this book:

- You can think of reproducibility as a form of robustness (see Chapter 11): A reproducible model is robust to the computing environment and to the random number generation.
- A lack of reproducibility introduces uncertainty (see Chapter 12). Let's say you forgot to set a random seed. This means that the next time the model is trained, it will be subject to uncertainty due to random splitting of the data and other operations based on a random number generator.
- Publication (see Chapter 14) and reproducibility both rely on documentation and go well together: If you have good reporting, you have also made your project more reproducible.

14. Reporting

If you don't publish the results of your experiments, can it still be considered science? If you don't share your results, others can't verify or build on your work. That's why publishing is so important in science. When you use machine learning in your research, there are additional needs for reporting. You may publish a paper but also report on the model or dataset you created. To make reporting easier, you can use checklists. Rather than just copying these checklists here – which would make for a terrible read – this chapter points you to the best checklists, their scope, and the ideas behind them.

After some years, other birds became interested in Rattle and Krarah's tornado prediction model. The pigeons, the magpies, and, of course, the parrots wanted to repeat the Ravens' success. But they had so many questions, such as what data was used in training? Krarah's disciples were tired of answering the same questions over and over again. They decided to compile all the information about the model into one document and publish it.

14.1. Checklist for papers

If you use machine learning in your research project, what details should you include in your paper? One checklist you might want to use is REFORMS, which stands for "Reporting standards for machine learning-based science" [231]. The authors developed REFORMS based on literature reviews and a consensus of 19 researchers from different fields. REFORMS also includes guidelines on completing the checklist and identifying potential problems in your modeling approach. REFORMS is *field agnostic* – this means the checklist items are broadly applicable, but some may be missing depending on your needs.

The REFORMS checklist consists of 8 modules (categories) and a total of 32 items:

- **Study goals.** Example: *Motivation for the use of machine learning methods in the study*
- **Computational reproducibility.** Example: *README file which contains instructions for generating the results using the provided dataset and code*
- **Data quality.** Example: *Sample size and outcome frequencies*
- **Data preprocessing.** Example: *Identification of whether any samples are excluded with a rationale for why they are excluded*
- **Modeling.** Example: *Method for selecting the model(s) reported in the paper*
- **Data leakage.** Example: *Justification that each feature or input used in the model is legitimate for the task at hand and does not lead to leakage*
- **Metrics and uncertainty.** Example: *Justification for the choice of statistical tests (if used) and a check for the assumptions of the statistical test*
- **Generalizability and limitations.** Example: *Evidence of external validity*

Further resources:

- REFORMS Website[1]
- REFORMS Checklist (Word Document)[2]
- REFORMS Guidelines (Word Document)[3]
- REFORMS Paper[4]

Some papers, especially in the medical field may require additional attention:

- TRIPOD+AI [232] is a publication approach for clinical prediction models. It comes with a checklist[5].
- CLAIM for AI[6] provides a checklist for AI papers about medical imaging [233].

14.2. Model Cards

Besides publishing a paper, you may want to share the model. You could just upload the model somewhere (like Github) and call it a day. The problem: Other researchers and interested parties will have trouble working with your model. A standard README file might be too short to describe the model, and a research paper might be too long and in the wrong format. The solution: Create a special document with your model such as a Model Card. "Model Cards" by Google Research [234] refers to both the approach and the resulting documents. Model Cards is a checklist approach used by many institutions, including HuggingFace[7] a large model hub and community for machine learning.

The original Model Cards paper introduces 10 categories each with multiple details prompts:

- **Model Details.** Example: *Model date*

[1] https://reforms.cs.princeton.edu/
[2] https://reforms.cs.princeton.edu/checklist.docx
[3] https://reforms.cs.princeton.edu/guidelines.docx
[4] https://www.science.org/doi/epdf/10.1126/sciadv.adk3452
[5] https://www.tripod-statement.org/wp-content/uploads/2019/12/TRIPOD AI_checklist.pdf
[6] https://pubs.rsna.org/page/ai/claim
[7] https://huggingface.co

- **Intended Use.** Example: *Primary intended use*
- **Factors.** Example: *Evaluation Factors*
- **Metrics.** Example: *Decision Thresholds*
- **Evaluation Data.** Example: *Datasets*
- **Training Data**
- **Quantitative Analyses**
- **Ethical Considerations**
- **Caveats and Recommendations**

The Model Cards approach is constantly evolving. For example, the Model Cards Guidebook [235] and template (see below) contain new and renamed categories. Model Cards is a great starting point for model documentation, but you should always customize it to your needs.

Further resources:

- Model Card paper[8] [234]
- Model Card template[9]
- Model Cards Guidebook[10]

Again, you may have more specific documentation requirements. For example, medical devices using machine learning have special reporting requirements, see this guideline[11] (focus on Germany).

14.3. Datasheets for Datasets

Carefully built and curated datasets are at the heart of machine learning. A lot of time and effort can go into collecting and cleaning data. Sometimes years. To maximize your return on investment, you might publish the dataset so that others can use it and cite your work. But just like the model, the dataset will raise a lot of

[8] https://arxiv.org/abs/1810.03993
[9] https://github.com/huggingface/huggingface_hub/blob/main/src/huggingface_hub/templates/modelcard_template.md
[10] https://huggingface.co/docs/hub/en/model-card-guidebook
[11] https://github.com/johner-institut/ai-guideline/blob/master/Guideline-AI-Medical-Devices_EN.md

questions if it doesn't come with the right information. When it comes to documentation, it is not wise to reinvent the wheel but to stand on the shoulders of giants. In this case, "Datasheets for Datasets" [236] is a popular approach to "facilitate communication between datasets creators and consumers."

Datasheets for Datasets lists 57 questions in 7 categories:

- **Motivation.** Example: *For what purpose was the dataset created?*
- **Composition.** Example: *What data does each instance consist of?*
- **Collection Process.** Example: *How was the data associated with each instance acquired?*
- **Preprocessing/cleaning/labeling.** Example: *Is the software that was used to preprocess/clean/label the data available?*
- **Uses.** Example: *Has the dataset been used for any tasks already?*
- **Distribution.** Example: *How will the dataset be distributed (e.g., tarball on website, API, GitHub)?*
- **Maintenance.** Example: *How can the owner/curator/manager of the dataset be contacted (e.g., email address)?*

Resources:

- Datasheets paper (PDF)[12] [236]
- Datasheets template (Markdown)[13]
- Datasheets template (Latex/Overleaf)[14]

[12] https://dl.acm.org/doi/pdf/10.1145/3458723

[13] https://github.com/fau-masters-collected-works-cgarbin/datasheet-for-dataset-template/tree/master

[14] https://www.overleaf.com/latex/templates/datasheet-for-dataset-template/jgqyyzyprxth

14.4. Reporting relies on good practice

If you want to document your research project well, you need to follow good research practices that go hand in hand with other topics in this book:

- Documenting the research project, model, and dataset forces you to think about use cases and limitations, which is closely related to how well your model generalizes to different contexts (Chapter 7).
- Publishing results from your model may require model insights, for which you may need interpretability (Chapter 9).
- Sharing models and data is also an integral part of reproducibility (Chapter 13).

15. The Future of Science in the Age of Machine Learning

We have reached the end of our little journey. Remember how we started with examples of scientists using machine learning to gain insights? From there, we evaluated machine learning from the perspective of different scientific goals, such as prediction, control, explanation, and understanding. We argued that machine learning has great potential to become a mainstay of scientific problem-solving. However, bare-bones supervised machine learning does not do the trick – we need add-ons such as interpretability, uncertainty quantification, and the integration of causal knowledge, all of which we systematically introduced in dedicated chapters. One by one, we addressed how to "fix" the shortcomings of machine learning for science: Generalization, domain knowledge, interpretability, causality, robustness, uncertainty, reproducibility, and reporting.

In this final chapter, we speculate about the future of science in the age of machine learning. How will scientists integrate machine learning into their research? Will some sciences remain untouched by machine learning? What will be the role of future scientists in the face of machine learning automation?

Time for some speculation.

Many years later, an aging but happy Rattle invited the stony-faced Krarah into her giant nest to talk about old times. Since then, several scientific problems have been solved, many with the help of machine learning. So, was it all good? Their smart feathers begin to vibrate. Rattle received a targeted ad from a radical political movement, and Krarah received a warning of an approaching storm. Krarah winked at Rattle: "No! Not all of it, but some of it for sure."

15.1. No science without machine learning

In early 2023, we started brainstorming and conceptualizing this book. By this time, the increasing use of machine learning models in science was becoming apparent, even in fields we wouldn't have thought to find machine learning in, such as anthropology and theoretical physics. Around the same time, in November 2022 to be exact, ChatGPT was released. One of many large language models that could be used as chatbots and for many natural language tasks. Today, these models can even integrate other inputs such as images. These so-called *foundation models* have a lot of potential and we can already see an impact on science:

- Similar to how doctors make judgments, multi-modal foundation models allow for the integration of medical data from multiple sources [237].
- Chatbots are opening the door to personalized education with high-quality feedback at a large scale [238].

- Large language models are even changing the way scientists write [239].

Foundation models have opened up new types of problems for the application of machine learning in science. We believe that in the future all sciences – not just the natural and social sciences but also the humanities – will integrate machine learning into their research. Machine learning will even be used in disciplines that have been beyond quantitative treatment like literature sciences, history, and philosophy. We do not claim that this development will always be for the better.

But especially with large language models like ChatGPT in mind, what role is left for humans in science anyway?

15.2. No science without humans

For decades, we have heard stories of theory-free science or dreams of the automation of science [240]. And, indeed, some of the tasks that scientists spend most of their time on have changed:

- Scientists no longer have to solve numerical equations by hand.
- Sensors collect and transmit data automatically, and there are fewer instances where scientists need to collect the data themselves.
- Plotting data can be done digitally and doesn't require pen and paper.

Many other tasks are now changing with machine learning.

- Labeling wildlife images may no longer be an issue for future animal ecologists [7].
- Machine learning models can help physicists design quantum experiments [241].
- Designing proteins with desired functions will no longer require the intuition of biologists, but can rely on the power of big data models like AlphaFold [12].

But may foundation models like ChatGPT at some point suggest all the experiments we should run and steer the direction of science?

It is true that more and more tasks that used to be done by scientists can be automated. But that doesn't mean we are moving toward a science without scientists. If you look at history, the job profile of scientists has always changed. In Aristotelian times, scientists were often collectors of knowledge, writing and reading were essential skills. Scientists like Copernicus or Galileo had to perform precise calculations using their brains instead of CPUs and a piece of paper instead of computer memory. Physicists in Einstein's time couldn't enter the field without strong training in pure mathematics. And in many fields of modern science, scientists without basic skills in programming and data analysis skills have a hard time. Similarly, the profile of future scientists will require basic machine learning modeling skills.

Automating a task can free up time for other tasks. Potentially, it can free up time for scientists to study new phenomena, which requires a different set of skills. That is, if writing grants and reviewing and submitting papers doesn't eat up all the extra time – but that's another story. Scientists are constantly reinventing their roles – discovering new things seems to be the only job description that never expires. In the end, science is what scientists find interesting, and for something to be interesting, there has to be a being with goals and desires who finds it interesting. In fact, we could start talking about artificial general intelligence and machine consciousness right now, but we prefer not to...

15.3. From predictive tool to scientific method

To become part of the scientific toolkit – as programming and data analysis already are today – machine learning skills must go beyond building a powerful predictive model. The perception of machine learning as a purely associative, predictive tool reflects a growing phase. As we have shown in this book, many other skills must be part of the scientist's machine learning toolkit to work for science:

- Nobody just gives you a prediction task, you need domain knowledge to identify a meaningful problem.
- Data is not given, it is constructed. Understanding your data is necessary to build robust and generalizing models.
- Uncertainties of predictions and causal dependencies are not a side quest but essential, especially when machine learning predictions guide real-world actions.
- Models that are complete black boxes, not reproducible, and poorly reported will not be used in practice.

Machine learning methodology needs to mature and take into account all aspects of science: We need best practices for formalizing scientific questions; frameworks for data acquisition, cleaning, and engineering; a theory of learning algorithms that explains why some learning algorithms are successful; and, finally, standards for evaluating deployed models.

We are optimistic that future machine learning will take these steps, otherwise we would not have written this book. Our optimism is supported by the fact that other scientific methods have followed a similar path. For example, early statistics was primarily a tool for summarizing demographic data or making questionable claims, as in phrenology, before it turned into the multi-faceted and widely accepted scientific methodology it is today.

15.4. Machine learning knowledge with a long shelf-life

Machine learning is a fast-moving field. New trends and hype every year can make it difficult to keep track of the topics that are here to stay. Specific algorithms such as random forests, support vector machines, convolutional neural networks, generative adversarial networks, graph neural networks, or transformer models always have a half-life. They may be state of the art for two to five years before the next competitor comes along.

How do you succeed in such a fast-paced environment? Of course, sometimes you have to learn a new tool. But ideally, you also learn the principles that remain. Learning to program involves learning both general concepts like functions and the specific syntax of a programming language. The concepts often live for a long time: Someone who learned about functions 30 years ago still finds that knowledge relevant today, but they may no longer write their applications in Delphi or Pascal.

What remains relevant in science is not the concrete methods for solving problems, but the problems themselves and the general strategies for solving them. This is what we tried to convey in this book. We introduced ideas and formalisms for describing questions as machine learning problems and provided taxonomies for solutions.

Ultimately, the future of science in the age of machine learning is not about replacing scientists but, about enhancing their ability to explore the unknown. And while the tools may change, the curiosity and drive that fuel scientific discovery will remain timeless. That, after all, is the essence of science: to keep asking, exploring, and understanding.

Acknowledgments

Writing this book was super fun. But it was a lot of work and we are very thankful for all the support we received.

Thanks to all the beta readers for their valuable feedback on our drafts: Diabb Zegpi Delgado, Gunnar König, Heidi Seibold, Ida Ahmad, Martine Jansen, Susanne Dandl, Thomas Freiesleben, Thomas Grote, Tom Sterkenburg, and Zelimir Kurtanjek.

Raven images for the cover and for the raven stories were created by jeeshiu from Fiverr[1].

Timo's postdoc position is funded by the Carl Zeiss Foundation project "Certification and Foundations of Safe Machine Learning Systems in Healthcare".

[1] https://www.fiverr.com/jeeshiu

Citing this Book

If you found this book useful for your research article, blog post, or product, we would be grateful if you would cite this book. You can cite the book like this:

```
Freiesleben, T. & Molnar, C. (2024).
Supervised Machine Learning for Science:
How to stop worrying and love your black box.
https://ml-science-book.com/
```

Or use the following BibTex entry:

```
@book{freiesleben2024supervised,
  title    = {Supervised Machine Learning for Science},
  author   = {Timo Freiesleben and Christoph Molnar},
  year     = {2024},
  subtitle = {How to stop worrying and love your black box},
  url      = {https://ml-science-book.com/}
}
```

We are always interested in how people use machine learning in their daily scientific work. If you use our book as a reference, it would be great if you wrote us a line and told us what for. This is of course optional and only serves to satisfy our curiosity and to stimulate interesting exchanges. Our email is science-book@christophmolnar.com

About the Authors

Christoph Molnar is interested in everything to do with machine learning. He is an independent researcher and book author based in Munich, Germany. Before this book, Christoph Molnar has written several other books[2]: Interpretable Machine Learning, Modeling Mindsets, Introduction To Conformal Prediction With Python, and Interpreting Machine Learning Models With SHAP. He also writes a newsletter called Mindful Modeler[3] about machine learning topics beyond pure prediction.

Before Christoph became an independent author, he did his PhD on interpretable machine learning at LMU Munich with Timo. Between his PhD and his Bachelors & Masters in Statistics at LMU, Christoph worked several years as a statistician and data scientist. In his spare time, Christoph enjoys cycling & calisthenics, shopping for groceries, cooking & eating delicious food, reading sci-fi novels by Liu Cixin & qntm, and ice bathing. You'll find him on Twitter[4], LinkedIn[5], and his personal website[6].

Timo Freiesleben is genuinely curious about machine learning and the philosophical questions it raises. Timo is a strange mix of a philosopher of science and a machine learning researcher. He is currently a post-doc at the "Machine Learning for Science" cluster at the University of Tübingen. Here are links to his website[7] and his Google Scholar profile[8]. In his spare time, Timo enjoys hiking, cooking, reading strange things like Dostoyevsky & Kafka,

[2] https://christophmolnar.com/books/
[3] https://mindfulmodeler.substack.com/
[4] https://twitter.com/ChristophMolnar
[5] https://www.linkedin.com/in/christoph-molnar/
[6] https://christophmolnar.com/
[7] https://freiesleben.github.io/
[8] https://scholar.google.com.au/citations?user=BUvMXaAAAAAJ&hl=de

doing creative things like drawing & poetic writing, sports like football, skating, & stand-up-paddling, and, like most people, spending quality time with friends & family.

Before his current position, Timo did his PhD with Christoph at LMU Munich, working mainly on interpretable machine learning, scientific modeling, and causality. He studied philosophy of science, mathematics, machine learning, and neuroscience at LMU Munich and the University of Tübingen. Timo is currently part of the Carl Zeiss Foundation project "Certification and Foundations of Safe Machine Learning Systems in Healthcare," which also funds his postdoc position.

References

[1] M. Reichstein *et al.*, "Deep learning and process understanding for data-driven earth system science," *Nature*, vol. 566, no. 7743, pp. 195–204, 2019, doi: 10.1038/s41586-019-0912-1.

[2] J. Schmidt, M. R. Marques, S. Botti, and M. A. Marques, "Recent advances and applications of machine learning in solid-state materials science," *npj Computational Materials*, vol. 5, no. 1, pp. 1–36, 2019, doi: 10.1038/s41524-019-0221-0.

[3] A. J. Kell, D. L. Yamins, E. N. Shook, S. V. Norman-Haignere, and J. H. McDermott, "A task-optimized neural network replicates human auditory behavior, predicts brain responses, and reveals a cortical processing hierarchy," *Neuron*, vol. 98, no. 3, pp. 630–644, 2018, doi: 10.1016/j.neuron.2018.03.044.

[4] M. A. Pedersen, "Editorial introduction: Towards a machinic anthropology," *Big Data & Society*, vol. 10. SAGE Publications Sage UK: London, England, p. 20539517231153803, 2023. doi: 10.1177/20539517231153803.

[5] J. Kim and V. Pavlovic, "Ancient coin recognition based on spatial coding," in *2014 22nd international conference on pattern recognition*, 2014, pp. 321–326. doi: 10.1109/ICPR.2014.64.

[6] Y.-H. He, "Machine-learning the string landscape," *Physics Letters B*, vol. 774, pp. 564–568, 2017, doi: 10.1016/j.physletb.2017.10.024.

[7] M. S. Norouzzadeh *et al.*, "Automatically identifying, counting, and describing wild animals in camera-trap images with deep learning," *Proceedings of the National Academy of Sciences*, vol. 115, no. 25, pp. E5716–E5725, 2018, doi: 10.1073/pnas.1719367115.

[8] R. Lagerquist, A. McGovern, C. R. Homeyer, D. J. Gagne II, and T. Smith, "Deep learning on three-dimensional multiscale data for next-hour tornado prediction," *Monthly Weather Review*, vol. 148, no. 7, pp. 2837–2861, 2020, doi: 10.1175/MWR-D-19-0372.1.

[9] Z. Zhang, Y. Jin, B. Chen, and P. Brown, "California almond yield prediction at the orchard level with a machine learning approach," *Frontiers in plant science*, vol. 10, p. 809, 2019, doi: 10.3389/fpls.2019.00809/full.

[10] T. C. Lucas, "A translucent box: Interpretable machine learning in ecology," *Ecological Monographs*, vol. 90, no. 4, p. e01422, 2020, doi: 10.1002/ecm.1422.

[11] G. L. Perry, R. Seidl, A. M. Bellvé, and W. Rammer, "An outlook for deep learning in ecosystem science," *Ecosystems*, pp. 1–19, 2022, doi: 10.1007/s10021-022-00789-y.

[12] J. Jumper *et al.*, "Highly accurate protein structure prediction with AlphaFold," *Nature*, vol. 596, no. 7873, pp. 583–589, 2021, doi: 10.1038/s41586-021-03819-2.

[13] J. Pereira, A. J. Simpkin, M. D. Hartmann, D. J. Rigden, R. M. Keegan, and A. N. Lupas, "High-accuracy protein structure prediction in CASP14," *Proteins: Structure, Function, and Bioinformatics*, vol. 89, no. 12, pp. 1687–1699, 2021, doi: 10.1002/prot.26171.

[14] M. Gupta *et al.*, "CryoEM and AI reveal a structure of SARS-CoV-2 Nsp2, a multifunctional protein involved in key host processes." *Research Square*, 2021, doi: 10.1101/2021.05.10.443524.

[15] "AlphaFold DB website." 2023. Available: https://alphafold.ebi.ac.uk/

[16] J. Pearl, "The limitations of opaque learning machines," *Possible minds*, vol. 25, pp. 13–19, 2019.

[17] J. Pearl and D. Mackenzie, *The book of why: The new science of cause and effect.* Basic books, 2018.

[18] G. Marcus, "The next decade in AI: Four steps towards robust artificial intelligence," *arXiv preprint arXiv:2002.06177*, 2020, doi: 10.48550/arXiv.2002.06177.

[19] N. Chomsky, I. Roberts, and J. Watumull, "Noam chomsky: The false promise of ChatGPT," *The New York Times*, vol. 8, 2023.

[20] L. Breiman, *Classification and Regression Trees.* New York: Routledge, 2017. doi: 10.1201/9781315139470.

[21] S. Bates, T. Hastie, and R. Tibshirani, "Cross-validation: What does it estimate and how well does it do it?" *Journal of the American Statistical Association*, pp. 1–12, 2023, doi: 10.1080/01621459.2023.2197686.

[22] F. Hutter, L. Kotthoff, and J. Vanschoren, *Automated machine learning: Methods, systems, challenges.* Springer Nature, 2019. doi: 10.1007/978-3-030-05318-5.

[23] R. Frigg and S. Hartmann, "Models in Science," in *The Stanford encyclopedia of philosophy*, Spring 2020., E. N. Zalta, Ed., https://plato.stanford.edu/archives/spr2020/entries/models-science/; Metaphysics Research Lab, Stanford University, 2020.

[24] Z. Pu and E. Kalnay, "Numerical weather prediction basics: Models, numerical methods, and data assimilation," *Handbook of hydrometeorological ensemble forecasting*, pp. 67–97, 2019.

[25] H. E. Douglas, "Reintroducing prediction to explanation," *Philosophy of Science*, vol. 76, no. 4, pp. 444–463, 2009, doi: 10.1086/648111.

[26] G. Shmueli, "To Explain or to Predict?" *Statistical Science*, vol. 25, no. 3, pp. 289–310, 2010, doi: 10.1214/10-STS330.

[27] N. Taleb, "The black swan: Why don't we learn that we don't learn," *NY: Random House*, vol. 1145, 2005.

[28] T. Freiesleben and T. Grote, "Beyond generalization: A theory of robustness in machine learning," *Synthese*, vol. 202, no. 4, p. 109, 2023, doi: 10.1007/s11229-023-04334-9.

[29] K. Popper, *The logic of scientific discovery.* Routledge, 2005.

[30] A. Chakravartty, "Scientific Realism," in *The Stanford encyclopedia of philosophy*, Summer 2017., E. N. Zalta, Ed., https://plato.stanford.edu/archives/sum2017/entries/scientific-realism/; Metaphysics Research Lab, Stanford University, 2017.

[31] P. J. Haley and D. Soloway, "Extrapolation limitations of multilayer feedforward neural networks," in *[Proceedings 1992] IJCNN international joint conference on neural networks*, IEEE, 1992, pp. 25–30. doi: 10.1109/IJCNN.1992.227294.

[32] E. Barnard and L. Wessels, "Extrapolation and interpolation in neural network classifiers," *IEEE Control Systems Magazine*, vol. 12, no. 5, pp. 50–53, 1992, doi: 10.1109/37.158898.

[33] U. Hasson, S. A. Nastase, and A. Goldstein, "Direct fit to nature: An evolutionary perspective on biological and artificial neural networks," *Neuron*, vol. 105, no. 3, pp. 416–434, 2020, doi: 10.1016/j.neuron.2019.12.002.

[34] C. Szegedy *et al.*, "Intriguing properties of neural networks," *arXiv preprint arXiv:1312.6199*, 2013, doi: 10.48550/arXiv.1312.6199.

[35] R. Balestriero, J. Pesenti, and Y. LeCun, "Learning in high dimension always amounts to extrapolation," *arXiv preprint arXiv:2110.09485*, 2021.

[36] T. S. Kuhn, *The structure of scientific revolutions*, vol. 962. University of Chicago press Chicago, 1997.

[37] B. Dennis, J. M. Ponciano, M. L. Taper, and S. R. Lele, "Errors in statistical inference under model misspecification: Evidence, hypothesis testing, and AIC," *Frontiers in Ecology and Evolution*, vol. 7, p. 372, 2019, doi: 10.3389/fevo.2019.00372.

[38] K. Hornik, "Approximation capabilities of multilayer feedforward networks," *Neural networks*, vol. 4, no. 2, pp. 251–257, 1991, doi: 10.1016/0893-6080(91)90009-T.

[39] Y. Lu and J. Lu, "A universal approximation theorem of deep neural networks for expressing probability distributions," *Advances in neural information processing systems*, vol. 33, pp. 3094–3105, 2020, doi: 10.5555/3495724.3495984.

[40] Y. Gao *et al.*, "Machine learning based early warning system enables accurate mortality risk prediction for COVID-19," *Nature communications*, vol. 11, no. 1, p. 5033, 2020, doi: 10.5281/zenodo.3991113.

[41] B. J. Erickson, P. Korfiatis, Z. Akkus, and T. L. Kline, "Machine learning for medical imaging," *Radiographics*, vol. 37, no. 2, pp. 505–515, 2017, doi: 10.1148/rg.2017160130.

[42] X. Ren *et al.*, "Deep learning-based weather prediction: A survey," *Big Data Research*, vol. 23, p. 100178, 2021, doi: 10.1016/j.bdr.2020.100178.

[43] T. Oikarinen *et al.*, "Deep convolutional network for animal sound classification and source attribution using dual audio recordings," *The Journal of the Acoustical Society of America*, vol. 145, no. 2, pp. 654–662, 2019, doi: 10.1121/1.5097583.

[44] W. Hu *et al.*, "Open graph benchmark: Datasets for machine learning on graphs," *Advances in neural information processing systems*, vol. 33, pp. 22118–22133, 2020, doi: doi/10.5555/3495724.3497579.

[45] M. Neethu and R. Rajasree, "Sentiment analysis in twitter using machine learning techniques," in *2013 fourth international conference on computing, communications and networking technologies (ICCCNT)*, IEEE, 2013, pp. 1–5. doi: 10.1109/ICCCNT.2013.6726818.

[46] J. Earman, "Bayes or bust? A critical examination of bayesian confirmation theory." MIT Press, 1992.

[47] J. Q. Toledo-Marín, G. Fox, J. P. Sluka, and J. A. Glazier, "Deep learning approaches to surrogates for solving the diffusion equation for mechanistic real-world simulations," *Frontiers in Physiology*, vol. 12, p. 667828, 2021, doi: 10.3389/fphys.2021.667828.

[48] R. Lam *et al.*, "Learning skillful medium-range global weather forecasting," *Science*, p. eadi2336, 2023, doi: 10.1126/science.adi2336.

[49] H. W. De Regt, "Understanding, values, and the aims of science," *Philosophy of Science*, vol. 87, no. 5, pp. 921–932, 2020, doi: 10.1086/710520.

[50] K. Maaz *et al.*, *Bildung in deutschland 2022: Ein indikatorengestützter bericht mit einer analyse zum bildungspersonal.* wbv Publikation, 2022. doi: 10.3278/6001820hw.

[51] P. W. Holland, "Statistics and causal inference," *Journal of the American statistical Association*, vol. 81, no. 396, pp. 945–960, 1986, doi: 10.2307/2289064.

[52] T. Freiesleben, G. König, C. Molnar, and Á. Tejero-Cantero, "Scientific inference with interpretable machine learning: Analyzing models to learn about real-world phenomena," *Minds and Machines*, vol. 34, no. 3, p. 32, 2024, doi: 10.1007/s11023-024-09691-z.

[53] M. B. McDermott, S. Wang, N. Marinsek, R. Ranganath, L. Foschini, and M. Ghassemi, "Reproducibility in machine learning for health research: Still a ways to go," *Science Translational Medicine*, vol. 13, no. 586, p. eabb1655, 2021, doi: 10.1126/scitranslmed.abb1655.

[54] M. Hardt and B. Recht, *Patterns, predictions, and actions: Foundations of machine learning.* Princeton University Press, 2022.

[55] G. Cybenko, "Approximation by superpositions of a sigmoidal function," *Mathematics of control, signals and systems*, vol. 2, no. 4, pp. 303–314, 1989, doi: 10.1007/BF02551274.

[56] P. R. Halmos, *Measure theory*, vol. 18. Springer, 2013. doi: 10.1007/978-1-4684-9440-2.

[57] I. Goodfellow, Y. Bengio, and A. Courville, *Deep learning.* MIT press, 2016.

[58] M. Belkin, D. Hsu, S. Ma, and S. Mandal, "Reconciling modern machine-learning practice and the classical bias–variance trade-off," *Proceedings of the National Academy of Sciences of the United States of America*, vol. 116, no. 32, pp. 15849–15854, Aug. 2019, doi: 10.1073/pnas.1903070116.

[59] J. W. Rocks and P. Mehta, "Memorizing without overfitting: Bias, variance, and interpolation in overparameterized models," *Physical review research*, vol. 4, no. 1, p. 013201, 2022, doi: 10.1103/PhysRevResearch.4.013201.

[60] R. Schaeffer *et al.*, "Double Descent Demystified: Identifying, Interpreting & Ablating the Sources of a Deep Learning Puzzle." arXiv, Mar. 2023. doi: 10.48550/arXiv.2303.14151.

[61] A. Curth, A. Jeffares, and M. van der Schaar, "A u-turn on double descent: Rethinking parameter counting in statistical learning," *Advances in Neural Information Processing Systems*, vol. 36, 2024.

[62] J. Frankle and M. Carbin, "The Lottery Ticket Hypothesis: Finding Sparse, Trainable Neural Networks." arXiv, Mar. 2019. doi: 10.48550/arXiv.1803.03635.

[63] P. L. Bartlett, P. M. Long, G. Lugosi, and A. Tsigler, "Benign Overfitting in Linear Regression," *Proceedings of the National Academy of Sciences*, vol. 117, no. 48, pp. 30063–30070, Dec. 2020, doi: 10.1073/pnas.1907378117.

[64] S. L. Smith, B. Dherin, D. G. Barrett, and S. De, "On the origin of implicit regularization in stochastic gradient descent," *arXiv preprint arXiv:2101.12176*, 2021, doi: 10.48550/arXiv.2101.12176.

[65] P. Rajpurkar *et al.*, "CheXNet: Radiologist-Level Pneumonia Detection on Chest X-Rays with Deep Learning." arXiv, Dec. 2017. doi: 10.48550/arXiv.1711.05225.

[66] L. Wynants *et al.*, "Prediction models for diagnosis and prognosis of covid-19: Systematic review and critical appraisal," *BMJ (Clinical research ed.)*, vol. 369, p. m1328, Apr. 2020, doi: 10.1136/bmj.m1328.

[67] M. Roberts *et al.*, "Common pitfalls and recommendations for using machine learning to detect and prognosticate for COVID-19 using chest radiographs and CT scans," *Nature Machine Intelligence*, vol. 3, no. 3, pp. 199–217, 2021, doi: 10.1038/s42256-021-00307-0.

[68] "The California Almond." Accessed: Feb. 16, 2024. [Online]. Available: https://www.waterfordnut.com/almond.html

[69] L. H. Clemmensen and R. D. Kjærsgaard, "Data Representativity for Machine Learning and AI Systems." arXiv, Feb. 2023. Accessed: Feb. 07, 2024. [Online]. Available: http://arxiv.org/abs/2203.04706

[70] K. Chasalow and K. Levy, "Representativeness in Statistics, Politics, and Machine Learning," in *Proceedings of the 2021 ACM Conference on Fairness, Accountability, and Transparency*, in FAccT '21. New York, NY, USA: Association for Computing Machinery, Mar. 2021, pp. 77–89. doi: 10.1145/3442188.3445872.

[71] D. H. Wolpert, "The lack of a priori distinctions between learning algorithms," *Neural computation*, vol. 8, no. 7, pp. 1341–1390, 1996, doi: 10.1162/neco.1996.8.7.1341.

[72] S. Shalev-Shwartz and S. Ben-David, *Understanding machine learning: From theory to algorithms.* Cambridge university press, 2014. doi: 10.1017/CBO9781107298019.

[73] T. F. Sterkenburg and P. D. Grünwald, "The no-free-lunch theorems of supervised learning," *Synthese*, vol. 199, no. 3, pp. 9979–10015, 2021, doi: 10.1007/s11229-021-03233-1.

[74] P. Maurage, A. Heeren, and M. Pesenti, "Does chocolate consumption really boost nobel award chances? The peril of over-interpreting correlations in health studies," *The Journal of Nutrition*, vol. 143, no. 6, pp. 931–933, 2013, doi: 10.3945/jn.113.174813.

[75] I. London, "Encoding cyclical continuous features - 24-hour time," *Ian London's Blog.* Jul. 2016. Accessed: Sep. 27, 2023. [Online]. Available: https://ianlondon.github.io/posts/encoding-cyclical-features-24-hour-time/

[76] A. Nguyen and M. R. Martínez, "MonoNet: Towards Interpretable Models by Learning Monotonic Features." arXiv, Sep. 2019. doi: 10.48550/arXiv.1909.13611.

[77] B. Hofner, A. Mayr, N. Robinzonov, and M. Schmid, "Model-based boosting in r: A hands-on tutorial using the r package mboost," *Computational statistics*, vol. 29, pp. 3–35, 2014, doi: 10.1007/s00180-012-0382-5.

[78] J. Goschenhofer, F. M. Pfister, K. A. Yuksel, B. Bischl, U. Fietzek, and J. Thomas, "Wearable-based parkinson's disease severity monitoring using deep learning," in *Machine learning and knowledge discovery in databases: European conference, ECML PKDD 2019, würzburg, germany, september 16–20, 2019, proceedings, part III*, Springer, 2020, pp. 400–415. doi: 10.1007/978-3-030-46133-1_24.

[79] M. Raissi, P. Perdikaris, and G. E. Karniadakis, "Physics Informed Deep Learning (Part I): Data-driven Solutions of Nonlinear Partial Differential Equations." arXiv, Nov. 2017. doi: 10.48550/arXiv.1711.10561.

[80] M. Raissi, P. Perdikaris, and G. E. Karniadakis, "Physics Informed Deep Learning (Part II): Data-driven Discovery of Nonlinear Partial Differential Equations." arXiv, Nov. 2017. doi: 10.48550/arXiv.1711.10566.

[81] F. Pfisterer, S. Coors, J. Thomas, and B. Bischl, "Multi-objective automatic machine learning with autoxgboostmc," *arXiv preprint arXiv:1908.10796*, 2019, doi: 10.48550/arXiv.1908.10796.

[82] C. Molnar, G. Casalicchio, and B. Bischl, "Quantifying model complexity via functional decomposition for better post-hoc interpretability," in *Machine learning and knowledge discovery in databases: International workshops of ECML PKDD 2019, würzburg, germany, september 16–20, 2019, proceedings, part i*, Springer, 2020, pp. 193–204. doi: 10.1007/978-3-030-43823-4_17.

[83] S. Letzgus and K.-R. Müller, "An explainable AI framework for robust and transparent data-driven wind turbine power curve models," *Energy and AI*, p. 100328, Dec. 2023, doi: 10.1016/j.egyai.2023.100328.

[84] R. Van Noorden and J. M. Perkel, "AI and science: What 1,600 researchers think," *Nature*, vol. 621, no. 7980, pp. 672–675, Sep. 2023, doi: 10.1038/d41586-023-02980-0.

[85] M. Flora, C. Potvin, A. McGovern, and S. Handler, "Comparing Explanation Methods for Traditional Machine Learning Models Part 1: An Overview of Current Methods and Quantifying Their Disagreement." arXiv, Nov. 2022. doi: 10.48550/arXiv.2211.08943.

[86] R. Roscher, B. Bohn, M. F. Duarte, and J. Garcke, "Explainable Machine Learning for Scientific Insights and Discoveries," *IEEE Access*, vol. 8, pp. 42200–42216, 2020, doi: 10.1109/ACCESS.2020.2976199.

[87] T. Miller, "Explanation in artificial intelligence: Insights from the social sciences," *Artificial Intelligence*, vol. 267, pp. 1–38, Feb. 2019, doi: 10.1016/j.artint.2018.07.007.

[88] F. Doshi-Velez and B. Kim, "Towards A Rigorous Science of Interpretable Machine Learning." arXiv, Mar. 2017. Accessed: Dec. 01, 2023. [Online]. Available: http://arxiv.org/abs/1702.08608

[89] A. Adadi and M. Berrada, "Peeking Inside the Black-Box: A Survey on Explainable Artificial Intelligence (XAI)," *IEEE Access*, vol. 6, pp. 52138–52160, 2018, doi: 10.1109/ACCESS.2018.2870052.

[90] B. Basso and L. Liu, "Seasonal crop yield forecast: Methods, applications, and accuracies," in *Advances in Agronomy*, vol. 154, Elsevier, 2019, pp. 201–255. doi: 10.1016/bs.agron.2018.11.002.

[91] J. H. Friedman and B. E. Popescu, "Predictive Learning via Rule Ensembles," *The Annals of Applied Statistics*, vol. 2, no. 3, pp. 916–954, 2008, doi: 10.1214/07-AOAS148.

[92] A. Zeileis, T. Hothorn, and K. Hornik, "Model-Based Recursive Partitioning," *Journal of Computational and Graphical Statistics*, vol. 17, no. 2, pp. 492–514, Jun. 2008, doi: 10.1198/106186008X319331.

[93] C. Rudin, C. Chen, Z. Chen, H. Huang, L. Semenova, and C. Zhong, "Interpretable machine learning: Fundamental principles and 10 grand challenges," *Statistics Surveys*, vol. 16, no. none, pp. 1–85, Jan. 2022, doi: 10.1214/21-SS133.

[94] Z. C. Lipton, "The Mythos of Model Interpretability." arXiv, Mar. 2017. doi: 10.48550/arXiv.1606.03490.

[95] L. Breiman, "Random Forests," *Machine Learning*, vol. 45, no. 1, pp. 5–32, Oct. 2001, doi: 10.1023/A:1010933404324.

[96] T. Freiesleben, "Artificial neural nets and the representation of human concepts," *arXiv preprint arXiv:2312.05337*, 2023, doi: 10.48550/arXiv.2312.05337.

[97] C. A. Scholbeck, C. Molnar, C. Heumann, B. Bischl, and G. Casalicchio, "Sampling, Intervention, Prediction, Aggregation: A Generalized Framework for Model-Agnostic Interpretations," in *Machine Learning and Knowledge Discovery in Databases*, P. Cellier and K. Driessens, Eds., in Communications in Computer and Information Science. Cham: Springer International Publishing, 2020, pp. 205–216. doi: 10.1007/978-3-030-43823-4_18.

[98] C. Molnar, *Interpretable machine learning: A guide for making black box models explainable*, 2nd ed. 2022. Available: https://christophm.github.io/interpretable-ml-book

[99] M. T. Ribeiro, S. Singh, and C. Guestrin, ""Why Should I Trust You?": Explaining the Predictions of Any Classifier," in *Proceedings of the 22nd ACM SIGKDD International Conference on Knowledge Discovery and Data Mining*, in KDD '16. New York, NY, USA: Association for Computing Machinery, Aug. 2016, pp. 1135–1144. doi: 10.1145/2939672.2939778.

[100] M. T. Ribeiro, S. Singh, and C. Guestrin, "Anchors: High-Precision Model-Agnostic Explanations," *Proceedings of the AAAI Conference on Artificial Intelligence*, vol. 32, no. 1, Apr. 2018, doi: 10.1609/aaai.v32i1.11491.

[101] S. Wachter, B. Mittelstadt, and C. Russell, "Counterfactual Explanations Without Opening the Black Box: Automated Decisions and the GDPR," *SSRN Electronic Journal*, 2017, doi: 10.2139/ssrn.3063289.

[102] E. Štrumbelj and I. Kononenko, "Explaining prediction models and individual predictions with feature contributions," *Knowledge and Information Systems*, vol. 41, no. 3, pp. 647–665, Dec. 2014, doi: 10.1007/s10115-013-0679-x.

[103] S. M. Lundberg and S.-I. Lee, "A unified approach to interpreting model predictions," in *Proceedings of the 31st International Conference on Neural Information Processing Systems*, in NIPS'17. Red Hook, NY, USA: Curran Associates Inc., Dec. 2017, pp. 4768–4777. doi: 10.5555/3295222.3295230.

[104] A. Goldstein, A. Kapelner, J. Bleich, and E. Pitkin, "Peeking Inside the Black Box: Visualizing Statistical Learning With Plots of Individual Conditional Expectation," *Journal of Computational and Graphical Statistics*, vol. 24, no. 1, pp. 44–65, Jan. 2015, doi: 10.1080/10618600.2014.907095.

[105] A. Fisher, C. Rudin, and F. Dominici, "All Models are Wrong, but Many are Useful: Learning a Variable's Importance by Studying an Entire Class of Prediction Models Simultaneously," *Journal of machine learning research : JMLR*, vol. 20, p. 177, 2019, Accessed: Jan. 16, 2024. [Online]. Available: https://www.ncbi.nlm.nih.gov/pmc/articles/PMC8323609/

[106] I. C. Covert, S. Lundberg, and S.-I. Lee, "Understanding global feature contributions with additive importance measures," in *Proceedings of the 34th International Conference on Neural Information Processing Systems*, in NIPS'20. Red Hook, NY, USA: Curran Associates Inc., Dec. 2020, pp. 17212–17223.

[107] J. H. Friedman, "Greedy function approximation: A gradient boosting machine." *The Annals of Statistics*, vol. 29, no. 5, pp. 1189–1232, Oct. 2001, doi: 10.1214/aos/1013203451.

[108] D. W. Apley and J. Zhu, "Visualizing the Effects of Predictor Variables in Black Box Supervised Learning Models," *Journal of the Royal Statistical Society Series B: Statistical Methodology*, vol. 82, no. 4, pp. 1059–1086, Sep. 2020, doi: 10.1111/rssb.12377.

[109] V. N. Vapnik, "An overview of statistical learning theory," *IEEE transactions on neural networks*, vol. 10, no. 5, pp. 988–999, 1999, doi: 10.1109/72.788640.

[110] C. Molnar *et al.*, "Relating the Partial Dependence Plot and Permutation Feature Importance to the Data Generating Process," in *Explainable Artificial Intelligence*, L. Longo, Ed., in Communications in Computer and Information Science. Cham: Springer Nature Switzerland, 2023, pp. 456–479. doi: 10.1007/978-3-031-44064-9_24.

[111] D. S. Watson and M. N. Wright, "Testing conditional independence in supervised learning algorithms," *Machine Learning*, vol. 110, no. 8, pp. 2107–2129, Aug. 2021, doi: 10.1007/s10994-021-06030-6.

[112] K. Aas, M. Jullum, and A. Løland, "Explaining individual predictions when features are dependent: More accurate approximations to Shapley values," *Artificial Intelligence*, vol. 298, p. 103502, Sep. 2021, doi: 10.1016/j.artint.2021.103502.

[113] C. Molnar, G. König, B. Bischl, and G. Casalicchio, "Model-agnostic Feature Importance and Effects with Dependent Features – A Conditional Subgroup Approach," *Data Mining and Knowledge Discovery*, Jan. 2023, doi: 10.1007/s10618-022-00901-9.

[114] J. Lei, M. G'Sell, A. Rinaldo, R. J. Tibshirani, and L. Wasserman, "Distribution-Free Predictive Inference for Regression," *Journal of the American Statistical Association*, vol. 113, no. 523, pp. 1094–1111, Jul. 2018, doi: 10.1080/01621459.2017.1307116.

[115] J. Howard *et al.*, "An evidence review of face masks against COVID-19," *Proceedings of the National Academy of Sciences*, vol. 118, no. 4, p. e2014564118, 2021, doi: 10.1073/pnas.2014564118.

[116] Y. Alimohamadi, M. Sepandi, M. Taghdir, and H. Hosamirudsari, "Determine the most common clinical symptoms in COVID-19 patients: A systematic review and meta-analysis," *Journal of preventive medicine and hygiene*, vol. 61, no. 3, p. E304, 2020, doi: 10.15167/2421-4248/jpmh2020.61.3.1530.

[117] O. D'Ecclesiis *et al.*, "Vitamin d and SARS-CoV2 infection, severity and mortality: A systematic review and meta-analysis," *PLoS One*, vol. 17, no. 7, p. e0268396, 2022, doi: 10.1371/journal.pone.0268396.

[118] F. Ayoub, T. Sato, and A. Sakuraba, "Football and COVID-19 risk: Correlation is not causation," *Clinical Microbiology and Infection*, vol. 27, no. 2, pp. 291–292, 2021, doi: 10.1016/j.cmi.2020.08.034.

[119] B. Schölkopf, D. Janzing, J. Peters, E. Sgouritsa, K. Zhang, and J. Mooij, "On causal and anticausal learning," *arXiv preprint arXiv:1206.6471*, 2012, doi: 10.48550/arXiv.1206.6471.

[120] Y. Gong and G. Zhao, "Wealth, health, and beyond: Is COVID-19 less likely to spread in rich neighborhoods?" *Plos one*, vol. 17, no. 5, p. e0267487, 2022, doi: 10.1371%2Fjournal.pone.0267487.

[121] C. Hitchcock and M. Rédei, "Reichenbach's Common Cause Principle," in *The Stanford encyclopedia of philosophy*, Summer 2021., E. N. Zalta, Ed., https://plato.stanford.edu/archives/sum2021/entries/physics-Rpcc/; Metaphysics Research Lab, Stanford University, 2021.

[122] T. Vigen, *Spurious correlations.* Hachette UK, 2015.

[123] J. Ellenberg, *How not to be wrong: The hidden maths of everyday life.* Penguin UK, 2014.

[124] J. B. Gibbons *et al.*, "Association between vitamin d supplementation and COVID-19 infection and mortality," *Scientific Reports*, vol. 12, no. 1, p. 19397, 2022, doi: 10.1038/s41598-022-24053-4.

[125] K. Yu *et al.*, "Causality-based feature selection: Methods and evaluations," *ACM Computing Surveys (CSUR)*, vol. 53, no. 5, pp. 1–36, 2020, doi: 10.1145/3409382.

[126] J. Adebayo, J. Gilmer, M. Muelly, I. Goodfellow, M. Hardt, and B. Kim, "Sanity checks for saliency maps," *Advances in neural information processing systems*, vol. 31, 2018, doi: 10.1609/aaai.v34i04.6064.

[127] S. Hu *et al.*, "Weakly supervised deep learning for covid-19 infection detection and classification from ct images," *IEEE Access*, vol. 8, pp. 118869–118883, 2020, doi: 10.1109/access.2020.3005510.

[128] A. Fischer, "Resistance of children to covid-19. how?" *Mucosal Immunology*, vol. 13, no. 4, pp. 563–565, 2020, doi: 10.1038/s41385-020-0303-9.

[129] J. Pearl, *Causality.* Cambridge university press, 2009.

[130] A. Sharma and E. Kiciman, "DoWhy: An end-to-end library for causal inference," *arXiv preprint arXiv:2011.04216*, 2020.

[131] J. Yoon, J. Jordon, and M. Van Der Schaar, "GANITE: Estimation of individualized treatment effects using generative adversarial nets," in *International conference on learning representations*, 2018.

[132] S. Athey, J. Tibshirani, and S. Wager, "Generalized random forests," 2019, doi: 10.1214/18-aos1709.

[133] M. C. Knaus, "Double machine learning-based programme evaluation under unconfoundedness," *The Econometrics Journal*, vol. 25, no. 3, pp. 602–627, Jun. 2022, doi: 10.1093/ectj/utac015.

[134] S. R. Künzel, J. S. Sekhon, P. J. Bickel, and B. Yu, "Metalearners for estimating heterogeneous treatment effects using machine learning," *Proceedings of the national academy of sciences*, vol. 116, no. 10, pp. 4156–4165, 2019, doi: 10.1073/pnas.1804597116.

[135] V. Chernozhukov *et al.*, "Double/debiased machine learning for treatment and structural parameters." Oxford University Press Oxford, UK, 2018. doi: 10.1111/ectj.12097.

[136] Bach, V. Chernozhukov, M. S. Kurz, and M. Spindler, "DoubleML – An object-oriented implementation of double machine learning in R." 2021. doi: 10.32614/cran.package.doubleml.

[137] Bach, V. Chernozhukov, M. S. Kurz, and M. Spindler, "DoubleML – An object-oriented implementation of double machine learning in Python," *Journal of Machine Learning Research*, vol. 23, no. 53, pp. 1–6, 2022, doi: 10.18637/jss.v108.i03.

[138] S. Dandl, "Causality concepts in machine learning: Heterogeneous treatment effect estimation with machine learning & model interpretation with counterfactual and semi-factual explanations," PhD thesis, lmu, 2023. doi: 10.5282/edoc.32947.

[139] J. Rothfuss, F. Ferreira, S. Walther, and M. Ulrich, "Conditional density estimation with neural networks: Best practices and benchmarks," *arXiv preprint arXiv:1903.00954*, 2019, doi: 10.48550/arXiv.1903.00954.

[140] S. Beckers and J. Vennekens, "A principled approach to defining actual causation," *Synthese*, vol. 195, no. 2, pp. 835–862, 2018, doi: 10.1007/s11229-016-1247-1.

[141] L. Jehi *et al.*, "Individualizing risk prediction for positive coronavirus disease 2019 testing: Results from 11,672 patients," *Chest*, vol. 158, no. 4, pp. 1364–1375, 2020, doi: 10.1016/j.chest.2020.05.580.

[142] G. König, T. Freiesleben, and M. Grosse-Wentrup, "Improvement-focused causal recourse (ICR)," in *Proceedings of the AAAI conference on artificial intelligence*, 2023, pp. 11847–11855. doi: 10.1609/aaai.v37i10.26398.

[143] J. Peters, D. Janzing, and B. Schölkopf, *Elements of causal inference: Foundations and learning algorithms.* The MIT Press, 2017.

[144] K. Fukumizu, A. Gretton, X. Sun, and B. Schölkopf, "Kernel measures of conditional dependence," *Advances in neural information processing systems*, vol. 20, 2007.

[145] R. Sen, A. T. Suresh, K. Shanmugam, A. G. Dimakis, and S. Shakkottai, "Model-powered conditional independence test," *Advances in neural information processing systems*, vol. 30, 2017, doi: 10.5555/3294996.3295055.

[146] R. D. Shah and J. Peters, "The hardness of conditional independence testing and the generalised covariance measure," *The Annals of Statistics*, vol. 48, no. 3, pp. 1514–1538, 2020, doi: 10.1214/19-AOS1857.

[147] M. Kalisch, M. Mächler, D. Colombo, M. H. Maathuis, and P. Bühlmann, "Causal inference using graphical models with the r package pcalg," *Journal of statistical software*, vol. 47, pp. 1–26, 2012, doi: 10.18637/jss.v047.i11.

[148] P. Spirtes, C. N. Glymour, and R. Scheines, *Causation, prediction, and search.* MIT press, 2000. doi: 10.1007/978-1-4612-2748-9.

[149] F. Eberhardt, C. Glymour, and R. Scheines, "On the number of experiments sufficient and in the worst case necessary to identify all causal relations among n variables," in *Proceedings of the twenty-first conference on uncertainty in artificial intelligence*, in UAI'05. Arlington, Virginia, USA: AUAI Press, 2005, pp. 178–184.

[150] C. F. Manski, *Partial identification of probability distributions*, vol. 5. Springer, 2003. doi: 10.1007/b97478.

[151] C. Uhler, G. Raskutti, P. Bühlmann, and B. Yu, "Geometry of the faithfulness assumption in causal inference," *The Annals of Statistics*, pp. 436–463, 2013.

[152] B. Neal, "Introduction to causal inference," *Course Lecture Notes (draft)*, 2020.

[153] S. Beckers and J. Y. Halpern, "Abstracting causal models," in *Proceedings of the aaai conference on artificial intelligence*, 2019, pp. 2678–2685. doi: 10.1609/aaai.v33i01.33012678.

[154] P. W. Koh *et al.*, "Concept bottleneck models," in *International conference on machine learning*, PMLR, 2020, pp. 5338–5348.

[155] Y. Bengio, A. Courville, and P. Vincent, "Representation learning: A review and new perspectives," *IEEE Transactions on Pattern Analysis and Machine Intelligence*, vol. 35, no. 8, pp. 1798–1828, 2013, doi: 10.1109/TPAMI.2013.50.

[156] B. Schölkopf *et al.*, "Toward causal representation learning," *Proceedings of the IEEE*, vol. 109, no. 5, pp. 612–634, 2021, doi: 10.1109/JPROC.2021.3058954.

[157] B. Schölkopf, "Causality for machine learning," in *Probabilistic and causal inference: The works of judea pearl*, 2022, pp. 765–804.

[158] L. Lei and E. J. Candès, "Conformal inference of counterfactuals and individual treatment effects," *Journal of the Royal Statistical Society Series B: Statistical Methodology*, vol. 83, no. 5, pp. 911–938, 2021, doi: 10.1111/rssb.12445.

[159] A. Swanson, M. Kosmala, C. Lintott, R. Simpson, A. Smith, and C. Packer, "Snapshot serengeti, high-frequency annotated camera trap images of 40 mammalian species in an african savanna," *Scientific data*, vol. 2, no. 1, pp. 1–14, 2015, doi: 10.1038/sdata.2015.26.

[160] J. Yang, K. Zhou, Y. Li, and Z. Liu, "Generalized Out-of-Distribution Detection: A Survey," *International Journal of Computer Vision*, Jun. 2024, doi: 10.1007/s11263-024-02117-4.

[161] P. Chattopadhyay, R. Vedantam, R. R. Selvaraju, D. Batra, and D. Parikh, "Counting everyday objects in everyday scenes," in *Proceedings of the IEEE conference on computer vision and pattern recognition*, 2017, pp. 1135–1144. doi: 10.1109/cvpr.2017.471.

[162] L. Ruff *et al.*, "A unifying review of deep and shallow anomaly detection," *Proceedings of the IEEE*, vol. 109, no. 5, pp. 756–795, 2021, doi: 10.1109/JPROC.2021.3052449.

[163] R. Chalapathy and S. Chawla, "Deep learning for anomaly detection: A survey," *arXiv preprint arXiv:1901.03407*, 2019.

[164] T. Neupert, M. H. Fischer, E. Greplova, K. Choo, and M. M. Denner, "Introduction to machine learning for the sciences," *arXiv preprint arXiv:2102.04883*, 2021, doi: 10.48550/arXiv.2102.04883.

[165] T. Denouden, R. Salay, K. Czarnecki, V. Abdelzad, B. Phan, and S. Vernekar, "Improving reconstruction autoencoder out-of-distribution detection with mahalanobis distance," *arXiv preprint arXiv:1812.02765*, 2018.

[166] B. Settles, "Active learning literature survey," University of Wisconsin–Madison, Computer Sciences Technical Report 1648, 2009.

[167] P. Ren *et al.*, "A survey of deep active learning," *ACM computing surveys (CSUR)*, vol. 54, no. 9, pp. 1–40, 2021, doi: 10.1145/3472291.

[168] T. Danka and P. Horvath, "modAL: A modular active learning framework for python," *arXiv preprint arXiv:1805.00979*, 2018.

[169] Y.-P. Tang, G.-X. Li, and S.-J. Huang, "ALiPy: Active learning in python," *arXiv preprint arXiv:1901.03802*, 2019, doi: 10.48550/arXiv.1901.03802.

[170] G. Ciravegna, F. Precioso, A. Betti, K. Mottin, and M. Gori, "Knowledge-driven active learning," in *Joint european conference on machine learning and knowledge discovery in databases*, Springer, 2023, pp. 38–54. doi: 10.1007/978-3-031-43412-9_3.

[171] B. Ghai, Q. V. Liao, Y. Zhang, R. Bellamy, and K. Mueller, "Explainable active learning (xal) toward ai explanations as interfaces for machine teachers," *Proceedings of the ACM on Human-Computer Interaction*, vol. 4, no. CSCW3, pp. 1–28, 2021, doi: 10.1145/3432934.

[172] A. Mumuni and F. Mumuni, "Data augmentation: A comprehensive survey of modern approaches," *Array*, vol. 16, p. 100258, 2022, doi: 10.1016/j.array.2022.100258.

[173] C. Shorten and T. M. Khoshgoftaar, "A survey on image data augmentation for deep learning," *Journal of big data*, vol. 6, no. 1, pp. 1–48, 2019, doi: 10.1186/s40537-019-0197-0.

[174] S. Feng *et al.*, "A survey of data augmentation approaches for NLP," in *Findings of the association for computational linguistics: ACL-IJCNLP 2021*, Association for Computational Linguistics, 2021. doi: 10.18653/v1/2021.findings-acl.84.

[175] I. Goodfellow *et al.*, "Generative adversarial nets," *Advances in neural information processing systems*, vol. 27, 2014.

[176] A. Antoniou, A. Storkey, and H. Edwards, "Data augmentation generative adversarial networks," *arXiv preprint arXiv:1711.04340*, 2017.

[177] P. Isola, J.-Y. Zhu, T. Zhou, and A. A. Efros, "Image-to-image translation with conditional adversarial networks," in *Proceedings of the IEEE conference on computer vision and pattern recognition*, 2017, pp. 1125–1134. doi: 10.1109/CVPR.2017.632.

[178] M. D. Bloice, C. Stocker, and A. Holzinger, "Augmentor: An image augmentation library for machine learning," *Journal of Open Source Software*, vol. 2, no. 19, p. 432, 2017, doi: 10.21105/joss.00432.

[179] S. Bird, "NLTK: The natural language toolkit," in *Proceedings of the COLING/ACL 2006 interactive presentation sessions*, 2006, pp. 69–72. doi: 10.3115/1225403.1225421.

[180] D. Hendrycks *et al.*, "The many faces of robustness: A critical analysis of out-of-distribution generalization," in *Proceedings of the IEEE/CVF international conference on computer vision*, 2021, pp. 8340–8349. doi: 10.1109/ICCV48922.2021.00823.

[181] D. Tsipras, S. Santurkar, L. Engstrom, A. Turner, and A. Madry, "Robustness may be at odds with accuracy," *arXiv preprint arXiv:1805.12152*, 2018, doi: 10.48550/arXiv.1805.12152.

[182] S.-A. Rebuffi, S. Gowal, D. A. Calian, F. Stimberg, O. Wiles, and T. A. Mann, "Data augmentation can improve robustness," *Advances in Neural Information Processing Systems*, vol. 34, pp. 29935–29948, 2021, doi: 10.48550/arXiv.2111.05328.

[183] S. Alemohammad *et al.*, "Self-consuming generative models go mad," *arXiv preprint arXiv:2307.01850*, 2023.

[184] G. Chandrashekar and F. Sahin, "A survey on feature selection methods," *Computers & electrical engineering*, vol. 40, no. 1, pp. 16–28, 2014, doi: 10.1016/j.compeleceng.2013.11.024.

[185] C. Vens, J. Struyf, L. Schietgat, S. Džeroski, and H. Blockeel, "Decision trees for hierarchical multi-label classification," *Machine learning*, vol. 73, pp. 185–214, 2008, doi: 10.1007/s10994-008-5077-3.

[186] S. Wang *et al.*, "Defensive dropout for hardening deep neural networks under adversarial attacks," in *2018 IEEE/ACM international conference on computer-aided design (ICCAD)*, IEEE, 2018, pp. 1–8. doi: 10.1145/3240765.3264699.

[187] A. Mumuni and F. Mumuni, "CNN architectures for geometric transformation-invariant feature representation in computer vision: A review," *SN Computer Science*, vol. 2, no. 5, p. 340, 2021, doi: 10.1007/s42979-021-00735-0.

[188] G. Corso, H. Stark, S. Jegelka, T. Jaakkola, and R. Barzilay, "Graph neural networks," *Nature Reviews Methods Primers*, vol. 4, no. 1, p. 17, 2024.

[189] Z. Wu, S. Pan, F. Chen, G. Long, C. Zhang, and S. Y. Philip, "A comprehensive survey on graph neural networks," *IEEE transactions on neural networks and learning systems*, vol. 32, no. 1, pp. 4–24, 2020, doi: 10.1109/TNNLS.2020.2978386.

[190] P. W. Battaglia *et al.*, "Relational inductive biases, deep learning, and graph networks," *arXiv preprint arXiv:1806.01261*, 2018.

[191] M. Arjovsky, L. Bottou, I. Gulrajani, and D. Lopez-Paz, "Invariant risk minimization," *arXiv preprint arXiv:1907.02893*, 2019.

[192] P. Kamath, A. Tangella, D. Sutherland, and N. Srebro, "Does invariant risk minimization capture invariance?" in *International conference on artificial intelligence and statistics*, PMLR, 2021, pp. 4069–4077.

[193] K. Cranmer, J. Brehmer, and G. Louppe, "The frontier of simulation-based inference," *Proceedings of the National Academy of Sciences*, vol. 117, no. 48, pp. 30055–30062, 2020, doi: 10.1073/pnas.1912789117.

[194] H. Zhang, H. Chen, Z. Song, D. Boning, I. S. Dhillon, and C.-J. Hsieh, "The limitations of adversarial training and the blind-spot attack," *arXiv preprint arXiv:1901.04684*, 2019, doi: 10.48550/arXiv.1901.04684.

[195] I. J. Goodfellow, J. Shlens, and C. Szegedy, "Explaining and harnessing adversarial examples," *arXiv preprint arXiv:1412.6572*, 2014, doi: 10.48550/arXiv.1412.6572.

[196] T. Tanay and L. Griffin, "A boundary tilting persepective on the phenomenon of adversarial examples," *arXiv preprint arXiv:1608.07690*, 2016, doi: 10.48550/arXiv.1608.07690.

[197] S. Han, C. Lin, C. Shen, Q. Wang, and X. Guan, "Interpreting adversarial examples in deep learning: A review," *ACM Computing Surveys*, vol. 55, no. 14s, pp. 1–38, 2023, doi: 10.1145/3594869.

[198] N. Papernot, P. McDaniel, and I. Goodfellow, "Transferability in machine learning: From phenomena to black-box attacks using adversarial samples," *arXiv preprint arXiv:1605.07277*, 2016, doi: 10.48550/arXiv.1605.07277.

[199] A. Ilyas, S. Santurkar, D. Tsipras, L. Engstrom, B. Tran, and A. Madry, "Adversarial examples are not bugs, they are features," *Advances in neural information processing systems*, vol. 32, 2019.

[200] T. Freiesleben, "The intriguing relation between counterfactual explanations and adversarial examples," *Minds and Machines*, vol. 32, no. 1, pp. 77–109, 2022, doi: 10.1007/s11023-021-09580-9.

[201] C. Gruber, P. O. Schenk, M. Schierholz, F. Kreuter, and G. Kauermann, "Sources of Uncertainty in Machine Learning – A Statisticians' View." arXiv, May 2023. doi: 10.48550/arXiv.2305.16703.

[202] S. Hoffmann, F. Schönbrodt, R. Elsas, R. Wilson, U. Strasser, and A.-L. Boulesteix, “The multiplicity of analysis strategies jeopardizes replicability: Lessons learned across disciplines,” *Royal Society Open Science*, vol. 8, no. 4, p. 201925, 2021, doi: 10.1098/rsos.201925.

[203] E. Begoli, T. Bhattacharya, and D. Kusnezov, “The need for uncertainty quantification in machine-assisted medical decision making,” *Nature Machine Intelligence*, vol. 1, no. 1, pp. 20–23, 2019, doi: 10.1038/s42256-018-0004-1.

[204] E. Hüllermeier and W. Waegeman, “Aleatoric and epistemic uncertainty in machine learning: An introduction to concepts and methods,” *Machine Learning*, vol. 110, no. 3, pp. 457–506, Mar. 2021, doi: 10.1007/s10994-021-05946-3.

[205] C. List, “Levels: Descriptive, explanatory, and ontological,” *Noûs*, vol. 53, no. 4, pp. 852–883, 2019.

[206] M. J. Van der Laan and S. Rose, *Targeted learning*, vol. 1. Springer, 2011. doi: 10.1007/978-1-4419-9782-1.

[207] P. Domingos, “A unified bias-variance decomposition,” in *Proceedings of 17th international conference on machine learning*, Morgan Kaufmann Stanford, 2000, pp. 231–238.

[208] A. Diamantopoulos, P. Riefler, and K. P. Roth, “Advancing formative measurement models,” *Journal of business research*, vol. 61, no. 12, pp. 1203–1218, 2008, doi: 10.1016/j.jbusres.2008.01.009.

[209] D. Carpentras, “We urgently need a culture of multi-operationalization in psychological research,” *Communications Psychology*, vol. 2, no. 1, p. 32, 2024.

[210] T. Hastie, R. Tibshirani, J. H. Friedman, and J. H. Friedman, *The elements of statistical learning: Data mining, inference, and prediction*, vol. 2. Springer, 2009.

[211] B. Jalaian, M. Lee, and S. Russell, “Uncertain context: Uncertainty quantification in machine learning,” *AI Magazine*, vol. 40, no. 4, pp. 40–49, 2019, doi: 10.1609/aimag.v40i4.4812.

[212] Y. Gal, “Uncertainty in deep learning,” 2016.

[213] A. Kendall and Y. Gal, "What uncertainties do we need in bayesian deep learning for computer vision?" *Advances in neural information processing systems*, vol. 30, 2017.

[214] L. S. Gottfredson, "The general intelligence factor." Scientific American, Incorporated, 1998.

[215] C. Strobl, A.-L. Boulesteix, T. Kneib, T. Augustin, and A. Zeileis, "Conditional variable importance for random forests," *BMC bioinformatics*, vol. 9, pp. 1–11, 2008.

[216] L. Mentch and G. Hooker, "Quantifying uncertainty in random forests via confidence intervals and hypothesis tests," *Journal of Machine Learning Research*, vol. 17, no. 26, pp. 1–41, 2016.

[217] Y. Gal and Z. Ghahramani, "Dropout as a bayesian approximation: Representing model uncertainty in deep learning," in *International conference on machine learning*, PMLR, 2016, pp. 1050–1059.

[218] T. J. Diciccio and J. P. Romano, "A review of bootstrap confidence intervals," *Journal of the Royal Statistical Society Series B: Statistical Methodology*, vol. 50, no. 3, pp. 338–354, 1988.

[219] C. Nadeau and Y. Bengio, "Inference for the generalization error," *Advances in neural information processing systems*, vol. 12, 1999.

[220] L. Semenova, H. Chen, R. Parr, and C. Rudin, "A path to simpler models starts with noise," *Advances in neural information processing systems*, vol. 36, 2024.

[221] S. W. Fleming, V. V. Vesselinov, and A. G. Goodbody, "Augmenting geophysical interpretation of data-driven operational water supply forecast modeling for a western US river using a hybrid machine learning approach," *Journal of Hydrology*, vol. 597, p. 126327, 2021.

[222] I. Takeuchi, Q. V. Le, T. D. Sears, A. J. Smola, and C. Williams, "Nonparametric quantile estimation." *Journal of machine learning research*, vol. 7, no. 7, 2006.

[223] J. Platt *et al.*, "Probabilistic outputs for support vector machines and comparisons to regularized likelihood methods," *Advances in large margin classifiers*, vol. 10, no. 3, pp. 61–74, 1999.

[224] Y. Romano, E. Patterson, and E. Candes, "Conformalized quantile regression," *Advances in neural information processing systems*, vol. 32, 2019.

[225] Y. Romano, M. Sesia, and E. Candes, "Classification with valid and adaptive coverage," *Advances in Neural Information Processing Systems*, vol. 33, pp. 3581–3591, 2020.

[226] V. Vovk, G. Shafer, and I. Nouretdinov, "Self-calibrating probability forecasting," *Advances in neural information processing systems*, vol. 16, 2003.

[227] O. S. Collaboration, "Estimating the reproducibility of psychological science," *Science*, vol. 349, no. 6251, p. aac4716, 2015, doi: 10.1126/science.aac4716.

[228] B. Arnold *et al.*, "The turing way: A handbook for reproducible data science," *Zenodo*, 2019.

[229] H. Seibold, *6 steps towards reproducible research.* Zenodo, 2024. doi: 10.5281/zenodo.12744715.

[230] L. Bottou, "Large-Scale Machine Learning with Stochastic Gradient Descent," in *Proceedings of COMPSTAT'2010*, Y. Lechevallier and G. Saporta, Eds., Heidelberg: Physica-Verlag HD, 2010, pp. 177–186. doi: 10.1007/978-3-7908-2604-3_16.

[231] S. Kapoor *et al.*, "REFORMS: Consensus-based Recommendations for Machine-learning-based Science," *Science Advances*, vol. 10, no. 18, p. eadk3452, May 2024, doi: 10.1126/sciadv.adk3452.

[232] G. S. Collins *et al.*, "TRIPOD+AI statement: Updated guidance for reporting clinical prediction models that use regression or machine learning methods," *BMJ (Clinical research ed.)*, vol. 385, 2024, doi: 10.1136/bmj-2023-078378.

[233] A. S. Tejani *et al.*, "Checklist for Artificial Intelligence in Medical Imaging (CLAIM): 2024 Update," *Radiology: Artificial Intelligence*, vol. 6, no. 4, p. e240300, Jul. 2024, doi: 10.1148/ryai.240300.

[234] M. Mitchell *et al.*, "Model Cards for Model Reporting," in *Proceedings of the Conference on Fairness, Accountability, and Transparency*, in FAT* '19. New York, NY, USA: Association for Computing Machinery, Jan. 2019, pp. 220–229. doi: 10.1145/3287560.3287596.

[235] E. Ozoani, M. Gerchick, and M. Mitchell, *Model card guidebook.* Hugging Face, 2022. Available: https://huggingface.co/docs/hub/en/model-card-guidebook

[236] T. Gebru *et al.*, "Datasheets for datasets," *Communications of the ACM*, vol. 64, no. 12, pp. 86–92, 2021, doi: 10.1145/3458723.

[237] M. Moor *et al.*, "Foundation models for generalist medical artificial intelligence," *Nature*, vol. 616, no. 7956, pp. 259–265, 2023.

[238] E. Kasneci *et al.*, "ChatGPT for good? On opportunities and challenges of large language models for education," *Learning and individual differences*, vol. 103, p. 102274, 2023.

[239] D. Kobak, R. G. Márquez, E.-Á. Horvát, and J. Lause, "Delving into ChatGPT usage in academic writing through excess vocabulary," *arXiv preprint arXiv:2406.07016*, 2024.

[240] C. Anderson, "The end of theory: The data deluge makes the scientific method obsolete," *Wired magazine*, vol. 16, no. 7, pp. 16–07, 2008.

[241] M. Krenn, M. Malik, R. Fickler, R. Lapkiewicz, and A. Zeilinger, "Automated search for new quantum experiments," *Physical review letters*, vol. 116, no. 9, p. 090405, 2016.

www.ingramcontent.com/pod-product-compliance
Lightning Source LLC
Chambersburg PA
CBHW050523240926
51846CB00003B/669
9783911578004